EXPEDITIONS

Gold, Shamans, and Green Fire

A TRUE ADVENTURE

by Lee Elders

Published by Wakani North LLC

ISBN: 978-0-9889504-0-5

Library of Congress Cataloging-in-Publication Data
Elders, Lee.

Expeditions: Gold, Shamans, and Green Fire

Cover Design:
Sue Denniston, Sage Spider

Layout and Typesetting:
Erin Campbell

First Edition
Printed in the Unites States

LeeElders.com

To Brit, my wife, best friend and soul mate:
Thank you for being the woman-child you are.

Acknowlegements

It would have been impossible for me to take this journey into the Republic of Ecuador without the help and friendship of my partner and fellow explorer, Dr. Adriano Vintimilla. Many thanks to his mother and father for allowing their son to invite me into their home and allowing me to become a member of their family.

I'd like to single out Bob Olson who believed in the Little Hell project and me and came running to assist. I'm deeply appreciative to the Canari Indians and the village of Nabon that trusted and worked with me, and to the priest of Nabon for his faith in our journey.

Padre Carlos Crespi gave openly of his time, knowledge, and his discovery of the emerald laden matrix found in Rio Tutanangosa. His information confirmed another piece to the treasure's puzzle.

Antonio Necta, the pener uwisin shaman who took me under his wing and shared his knowledge of plants, wildlife and tribal tradition became a true friend. He was the point man for all my expeditions in the land of the Shuara and risked his life and power to help me reach my goal.

I value all of the Shuara Indians who were trustworthy and never complained about the heavy loads of food, tents and supplies they hauled through the jungle. We became brothers.

My thanks to my brother-in-law, Bill Wright shared my dreams and goals and with daring and courage led dangerous expeditions for our team, up the Zamora River and the headwaters of the Rio Tungus.

I owe a special thanks to my mother, Erlene Kennedy, who thought I was crazy for traveling such a different path of life, but who always supported my goals, and to Barbara and Harris Waren and Brett Nilsson who were my biggest cheering section.

Some of those mentioned in Expeditions that I crossed paths with or became friends with are no longer with us. I would like to pay special tribute to Jim Bell, Bob Olson, Padre Carlos Crespi, Ted Vintimilla and Mr. & Mrs. Vintimilla, to name a few. God speed to all of you!

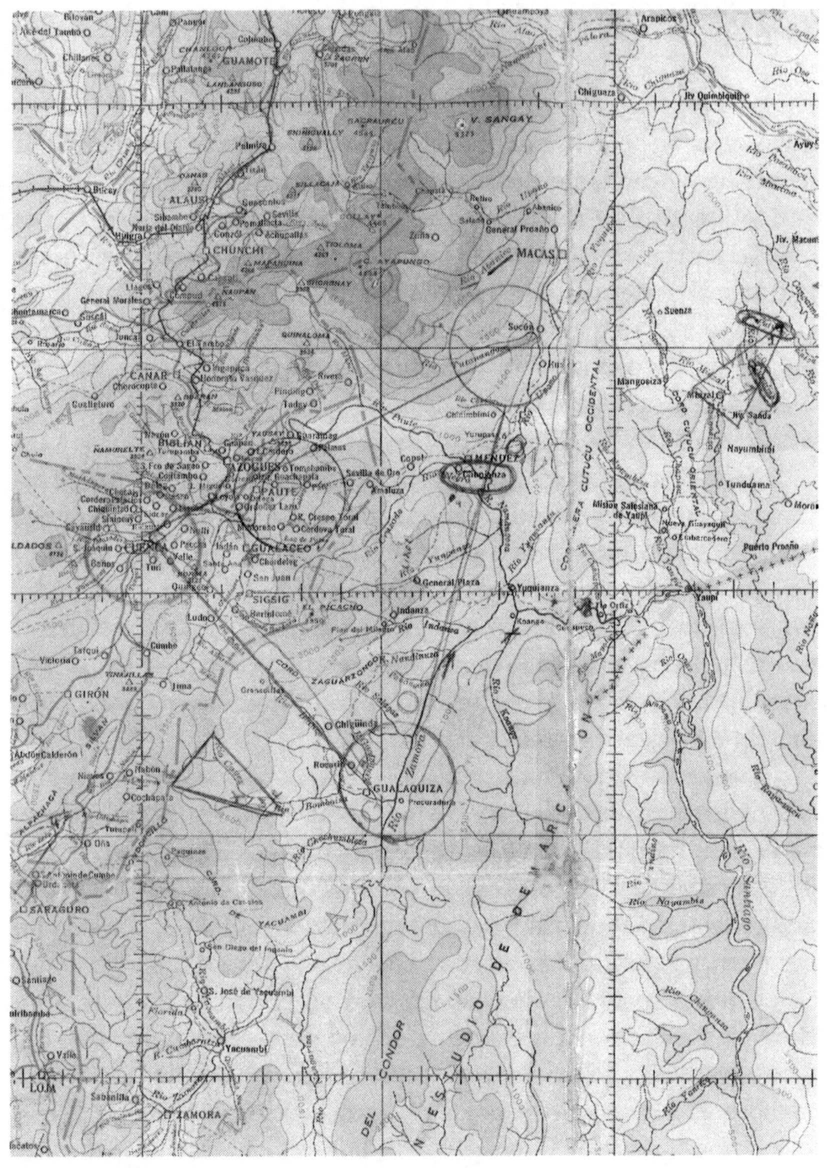

Table of Contents

INTRODUCTION

During the mid-sixties as the United States planned to place humans on the moon, the vast unexplored Oriente region of Ecuador with an infant mortality rate of forty-percent and a life expectancy in the forties had forgone any hope of entering the twentieth century. Although the land was unfathomably rich in natural resources of timber, gold and precious gems it was little consolation to those ragged souls mired in grinding poverty. It was indeed a land of contrasts where dense rain forest protected a lost emerald mine and spirit shrouded rivers their gold that became the magnetic pull for adventurers like me. But danger was always present and a constant reminder for an adventurer trapped between exploration amid the home of superstitious tribes, raging rivers, night stalkers and the ghosts of yesteryear.

Curses lingered in this wild lawless land. Some were hidden in tribal legend while others became more pronounced when outsiders entered their domain in search of personal wealth. Black magic consummated with present day curses permeated the super natural world of the Untsuri Shuara, the headhunters of the Western Amazon. The Shuara lived in a mystical realm of ayahuasca-induced visions with spirits that dictated life's path. The shamans' mysterious world was one where evil reined in two of three of their most prestigious positions. The Waweks, the bewitchers, the Kankaram, the killers and the Uwisin, the curers, were enveloped in a constant battle of good versus evil.

It seemed cruel that Creation would allow a Colombian, Rafael Mejia to play out his destiny in life by hiding a magnificent treasure of emeralds in a dense jungle referred to as the Kionade. Mejia's legacy would attract and compel ignorant souls to enter this no-mans land ruled by discarnate energy long passed but not forgotten.

Searching for Mejia's lost treasure required a journey into a land where time stood still, a place of legends and a fascinating ensemble of truth and reverence that translated from the lips of the unta Shuara as the place of the City of Magicians.

Undoubtedly, this mystical city was the birthplace of their tribal shamanism and they revered it as such. Even more intriguing, shamanism derives its name from the Tungus people of Siberia. Unexplainably the Rio Tungus was the mysterious name of the river that guarded the Kionade, separating the north bank from the Otherworld. Beneath the decoding of Mejia's 81 year-old riddle that pointed to the location of his lost emerald treasure and the slashing and hacking through dense jungle in search of his legacy; this book is also about an ethereal world encased within our third dimensional reality. However strange or incredible some of the following incidents may appear this is not a work of fiction.

Lee Elders
Pinewood, Arizona

CHAPTER 1

LITTLE HELL

Rio Infiernilios

Infiernilios was a ghost river. It wasn't found on Ecuadorian maps and defied definition in my Spanish dictionaries, yet was referred to by the locals as Little Hell, an ominous title that stirred the fire of the villager's souls. They said the river tumbled from the peak of the Cordillera Manga Hurco, winding through dense jungle and steep ravines until it reached the playa, the beach. There it pushed up sandbars and slowed to a crawl allowing golden tears of the sun to lodge in rocky crevices and glisten in the noonday sun. The gold nuggets found in the Infiernilios could change the desolate lives of the small, impoverished village of Nabon into one of promise of hope. But it was a dichotomy. It could transform yesterday's misery into tomorrow's promise of gold but it could also leave a trail of despondency. Rio Infiernilios was a combination of heaven and hell compressed into one small tributary that was a closely guarded secret of the inhabitants in the small village of Nabon.

Legend said that Manga Hurco, the towering mountain range that gave birth to Infiernilios, was a harbinger of human destiny, guarded by mountain spirits who carefully selected those who were permitted to cross her slopes. Only a few had been allowed the opportunity to wash the precious metal from the depths of Hell and from those few surfaced amazing stories that served to fuel the hearts and imaginations of those who dared not leave the sanctuary of Nabon.

Legend also said the mountain spirits were unforgiving to those whose purpose was motivated solely by greed and personal gain. It was said the Inca who worshiped the golden tears of the sun and fashioning the metal into sacred effigies never discovered the river. The history of the river indicated that the Conquistadors also had failed to locate the rich placer protected by the mountain spirits who kept their treasure secreted. No white man had ever set foot on the banks of Infiernilios; the colonials, as the fair-skinned ones are called, could not be trusted to share with the villagers.

The villagers of Nabon were superstitious, blanketed by ancient traditions, yet they were God-fearing parishioners who faithfully attended the old Catholic Church to pray for their sins and survival by asking God to find a way for them to be accepted by the mountain spirit guardians of the river. They knew they needed outside assistance; they felt incapable of pulling it off themselves. But, were they willing to divulge the secret to an outsider? Could that group or person be trusted? The local priest was sympathetic to their dilemma, but scolded them. "The spirit guardians you fear are nothing more or less than the spirit that resides in every one of you", he told them. "You are fearful of losing something that is not yours because it has never been yours to claim and own. The rivers and mountains are God's creation and they belong to him, no one else. But, God will share his wealth providing your hearts and minds are in the right place." He then instructed his parish to pray for guidance and not to forget that God worked mysteriously.

I'm not sure who or what I would credit, but life had moved me in mysterious ways. Months earlier the Rio Shingata had been responsible for my first trip into the Republic of Ecuador in the summer of 1966. Leaving behind the comfortable world of mortgage banking, I signed on as an underwater diver with an Arizona based gold dredging expedition. I had no diving experience, but I was willing to learn and they were willing to give me the opportunity. I saw it as the chance of a lifetime to return to the adventure and freedom of my childhood, and I looked forward to the challenge.

My employer had researched the gold bearing rivers of the Ecuadorian Andes and had settled on the Rio Shingata and they planned accordingly to maximize the return from the ore bearing waters. Listed on their equipment manifest was a Mack truck, a LSD conversion "Duck" needed to house the ten inch and 12 inch suction dredges, plus all the other equipment necessary to insure success in this type of alluvial underwater mining operation.

Everything had been carefully plotted and calculated, including the mining team, which was waiting on station at the Hotel Humboldt in the port city of Guayaquil. Success was at our fingertips and everyone was mentally spending the dollars they would convert from the gold. But, the visions of El Dorado were short lived. Ecuadorian customs refused to allow the equipment to be unloaded without a special tax of $50,000.00 U.S. dollars. My employers couldn't comprehend the Ecuadorian tradition of la mordita, which was nothing more than a payoff to the proper officials. The company considered the tax an insult and, on principle, refused to pay. They returned home, sadder but wiser.

Suddenly unemployed, I decided to stay in this mystifying country and strike out on my own. I knew the Rio Shingata was the targeted river and I knew its location. I decided to scout the area around the river to get a lay of the land. I was really rolling the dice. I was a foreigner in a strange land with limited ability in the Spanish language. I guess what I lacked in knowledge and language skills I certainly made up for in courage or naivety.

Getting out of the port city of Guayaquil was my first objective. I knew that the quickest way to access the Shingata was to base my operations in the central highlands city Cuenca, named for a bowl in a deep valley and also for a bend in a river. This would be my springboard for a journey to the small village tucked away from prying eyes along the Andean crest in the southern part of the Republic of Ecuador.

A man named Francisco had had an angel sitting on his shoulder that first day I arrived in Nabon. Religious proceedings had blocked my Land Rover on the side of a narrow mountain road and while the driver and I waited for it to pass, my driver spotted him and suggested that I hire the man carrying the cross on his back. "It would be a good omen," he said.

I studied Francisco, the loner, as he carried the burden of Christianity on his back for the local priest and the religious procession winding its way towards the sleepy village. When the procession was finished I hired the good omen. With that simple act Francisco regained his self-confidence. He had been jerked around his whole life by the demoralizing caste system of his culture and had escaped his demons through his fondness of drink. The former pariah of the village discovered a new sense of responsibility, curtailed his drinking and in time became one of my closest allies.

Camped on the banks of the Shingata, Francisco shared a secret that led me in another direction. Due east of his village, Nabon, was the Rio Infiernilios, the richest gold placer bearing river in the Andes. I sat for three hours in a downpour on the bank of the Shingata, listening intently to Francisco, who brought his friend Carlos in to reinforce his story about the legend of Infiernilios. Fumbling through my pocket Spanish dictionary and with the little Spanish I had picked up, a lengthy interrogation ensued. In an ugly display of Americanism and never one to mince words, I blurted out a challenge. "If it's so damned rich, why are you and the village so poor?"

Carlos, who had been to Little Hell on one occasion to wash the precious metal, replied quietly. "Senor Lee, the trip is long and dangerous and we have no tents, no sleeping bags and no backpacks, and no money to pay cargo carriers or buy extra food. I am one of the best batea men in my country. When I go to Infiernilios, I can stay only one day. It takes three days to go in and three days to come out. With blankets, machete, cooking pot and batea, I can only carry food for seven days."

With every question the men had rational answers. They explained that when they brought out gold, they were always paid far below the world price of $35.00 per ounce. Then I asked the big question. Why would they share their secret with a stranger, someone they barely knew, and a colonial no less? Their answer caught me off guard. "We been with you weeks and know you to be feria (fair) and honesto. Will you not hire poor people of village to work for you?" asked Francisco.

On the afternoon of the first day of the journey, seven men, one boy and three horses marched in single file across wind swept plains of nodding grass on the Andean crest. Finally, we reached the Rio Shingata, a winding tributary that also flowed from the cordillera Manga Hurco, a huge black-faced mountain to the south. We slumped into the tall weeds hoping to regain a little strength before committing our tired bodies to another round of torture. Propping myself on one elbow I gasped for air and stared at the swirling river coursing through this valley and wondered what an Arizona boy was doing in this water-logged wild country? This was a far cry from the flat open terrain I knew so well.

A freezing gust of wind brought me back to the present and the turbulent waters of the Rio Shingata, which we were about to navigate. We staggered to our numb feet and waded into the icy cold water as broiling black clouds moved southward. Bracing against the biting wind, Carlos sneered, "Senor Lee, damn cold river for me." Three of the Canari Indians started laughing at Carlos and his obvious struggle with a foreign language, which was phrased totally back-ass-wards from his own.

Chest high in cold currents, we broke into boisterous laughter caused by Carlos' mutilation of the language and partly to mask apprehension of the unknown that awaited us. Between the native's pigeon English and my mountain Spanish we were able to communicate in grunts, gestures and words that really never formed a complete sentence and seemed to border on Neanderthal. But, it had worked; our primitive communication established a bond between us and eliminated any unspoken racism in our quest for El Dorado.

Carlos was the true leader of my rag-tag team. He'd worked in construction on the Pan Am highway for a dollar a day, which was good pay for laborers. It was just enough money to put beans, yucca and bread on the table for his wife and two young sons. His appearance was more Spanish, unlike Francisco, who was Mestizo, or Ochoa and the others whose lineage was pure Canari. Carlos was interested in bettering his life. He was intelligent and creative but also ambitious, which was rare. He had not succumbed to self-pity, which was the breeding ground for alcoholism among the natives.

Ochoa was nicknamed the "parrot" by the other Canari and it was a name that suited him well. He was slender built with a bird-like face and never—and I mean never—stopped talking. I think he even talked in his sleep. But we were a team. We emerged from the icy water and for the next three hours continued our slow climb, knifing through low, racing clouds, bitterly cold winds, and rarified air that pained our lungs and left us dizzy and spent. When we reached the Rio Culebrillas, we pitched camp on a grassy knoll engulfed in a slow moving mist, which chilled us to the bone and made our struggle to erect and anchor tents more difficult in the milky, gray fog and gusty winds.

The young Canari boy that had accompanied his father's three horses for the first leg of the trip reminded me that he had to return home. I had agreed to pay 120 sucres, for the use of the horses and as I counted out the cash his eyes lit up. In American money this was equivalent to six dollars: two dollars, per horse, per day. We used the animals for only one day but I figured it was a bargain for both of us. The horses carried our tents, sleeping bags, shovels, bateas and a ten-day supply of food and water for our seven-man expedition. They were a pitiful lot and although sway backed and half starved, they still commanded a dollar a day more than the highest paid worker. Unfortunately, the caballos (horses) were only good for the first day, which was now finished. The second day the trail became too steep and muddy for secure footing, even for those poor beasts of burden.

As the boy excitedly tied the horses together I handed him a bag of jerky and shook his hand. I was concerned about him traveling through the night but Carlos assured me the Canari boys were not strangers to nights in the mountains. He waved good-bye and I had a flash of memory from my youth of the days and nights I spent alone in the Arizona desert. Youth has a way of convincing the young of their immortality. I knew that in the boy's mind he was a grown man headed on the trail back to Nabon. He knew that when he arrived his family would greet him with a hero's welcome for his first solo adventure and its success. I wished him well.

That evening, after the wind died down, I joined Francisco, Carlos, and Ochoa, in their tent to get our bearings for the next day's ordeal. They had already steamed up a bowl of aguardiente, which was Ecuador's version of White Lightning. When heated the sugarcane drink becomes extremely potent and the boys were well on their way to insulating themselves from the harshness of the first leg of our journey.

It was midnight when I crawled into my tent, to an encore of booming thunderclaps and crackling lightning. I lay there wondering what lay ahead before memories of my childhood flooded my mind, swirling in a montage of ghostly images of a bleak, harsh, desert and the emotional loneliness of growing up on the San Carlos Apache Indian Reservation in eastern Arizona.

In the days of my childhood, Apaches were truly third world citizens. They lived in hogans and teepees. These crude and primitive dwellings dotted the landscape around our old railroad house sitting next to a spur, a stone's throw away from the rusting rails of the Southern Pacific Railroad tracks. Peridot was a railroad siding, named from the pale green gemstone found in the mountains and was in the heart of the Apache Nation. It was similar to the other way stations my grandparents and I lived in during those hard times; places with strange sounding names like Dragoon, Black Water, Hyder, Sentinel and my birthplace Bowie, Arizona. Most could not be found on any map other than those used by the Southern Pacific railroad, but they existed and still lived in my heart.

My grandfather, Thomas Walter Kennedy began his employment on the Southern Pacific Railroad as a pick and shovel laborer at fifty-cents per day and eventually worked up the ranks to section foreman. Like the men he commanded, he, too, was once known in railroad jargon as a "gandy dancer" the unflattering term given to workers on a section gang. He could relate to those poor souls from Mexico and the reservation that were following his footsteps scratching out a living for themselves and their families. He told me time after time, "It wasn't much money, but I was damn glad to have it." They were a hearty bunch, those gandy dancers and the section foremen who toiled long hours in the boiling Arizona sun shoveling slag and laying rail.

Our clapboard railroad house lacked electricity and plumbing. It contained two bedrooms, a kitchen, living room, and a screened porch, which was our main sleeping area in the summer heat. That porch was a godsend that allowed cool breezes to help us get through the miserably hot summer nights. A small shed in back constructed from discarded railroad ties embalmed with and smelling of creosote, an oily liquid with the pungent odor of wood tar, greeted our senses each time we entered the makeshift bathroom and shower. The shed also doubled as a storage bin for the coal that fueled our cook stove in the kitchen. It was in this setting, at six years of age, I had my first experiences with danger and intrigue.

On Saturday nights the Apaches gathered in the San Carlos River bottom near our house in their timeless tradition of ceremonial dances and revelry. Their presence always seemed to coincide with the time my grandfather headed out for an evening of "honky-tonking" leaving my grandmother and I to fend for our selves. Somehow, the Apaches knew his departure patterns and the length of his expected absence, which offered them the opportunity to raid the coal shed. But, Granny, Minnie Lee Kennedy, was no fool, much less a shrinking violet. She was from sturdy pioneer stock and was always on the alert for those Apache bucks that crept through the dead of night without war paint, armed with a black bucket in each hand and sheer determination to steal our coal. Granny was a born again Christian who wouldn't hurt a fly. But, she would protect her coal.

The Apache bucks usually arrived around midnight after their ceremonies wound down to the final muted beat of the tom-toms along the banks of the dry river bottom. That was her signal to arm herself with Granddad's 32 Smith and Wesson. Crouched in the darkness of my little bedroom that looked down on the shed, she would place the unsuspecting coal poachers in her sights. I'd ask if she was going to shoot them. Her reply was always the same, "Only if things get out of hand." Fortunately for the Apache they always lived to raid our coal another night.

Some forty yards north of our house stood a metal water tower suspended on four fifty foot steel legs that housed our drinking and cooking water. It was an ugly structure that was cleaned periodically of dead birds that had somehow become caught in the chamber and were unable to escape. In back of the tower was what my grandfather referred to as "Indian quarters" a long barracks shaped building that, as far as I could remember, never served as home for Apache workers, but did serve as living quarters for hundreds of Mexican nationals that had been recruited to become steel driving men.

The brown skinned people wore large hats they called sombreros and spoke a strange language that drew my curiosity. I would sneak down to the quarters, hiding in the shadows, observing what Minnie Lee referred to as, "that army of heathens and banditos" as they prepared their evening meals and relaxed after a backbreaking day. It wasn't long before they detected my presence and instead of sacrificing me to their gods or robbing me blind, as I feared, they invited me into their cramped world to share their food and hospitality. I found myself enjoying their company and trading rudimentary English lessons for tortillas and frijoles. It was here, at the tender age of six that I learned my first words of Spanish and the true meaning of sharing and giving with people of different ethnic backgrounds.

I fell asleep reliving those memories and awoke to the smell of fresh coffee and a serious hangover. I crawled out of my tent, rubbed my sore legs, I noticed that all the other tents had already been struck and that Carlos, Ochoa, Francisco and the three cargo carriers were packed up and waiting for me to join them over a breakfast of pork, rice and beans. Their eagerness and my tardiness only enhanced my already shitty attitude. My head felt like it was compressed in a vice, my tongue was as dry as corn silk, and the foul taste in my mouth defied description.

I stared at Francisco, who was sitting on his backpack; he looked as bad as I felt. Something felt different. I tried to shake off the cobwebs in my brain and focus my eyes, I glanced around and then my eyes locked onto an unfamiliar addition to our group. Sitting at safe distance behind the fearless Canari was a brownish-yellow dog that had mysteriously joined our expedition. Carlos noticed my surprise and came forward to explain that the dog had joined up with us early that morning and that none of the boys knew anything about him. I named him Perro, which wasn't too original, because in Spanish perro means dog. He, like everything else in this region, was undernourished and eager to be accepted.

Francisco sidled up next to me and offered a slug of "Pelo de Perro," hair of the dog. I took a swig of the aguardiente and almost threw up. It was the same shit we had been drinking the night before; only it was cold. I grimaced as the bitter taste crossed my lips. While the crew laughed, Perro, our newfound friend was making short order of someone's left over rice and beans. Perro wasn't too trusting and kept one eye on the tin plate and the other on us.

The following morning, we paused briefly in the early morning sun, to survey the beautiful panoramic view of the narrow valley that lay below us. It was an incredible sight. A small river snaked its way through a sea of green grass and disappeared into the endless shadows cast by the towering cliffs to the north and south. Carlos had said that we would have to make an early camp, quite possibly by four in the afternoon. Seeing the canyon ahead I understood his logic; the Cordillera Yana Cocha, the massive mountain range to the south, would block the sun. He also said with some dismay that our camp for the night would be next to the Yana Cocha, an Inca phrase translated from Quichua, their native language that meant "Black Lagoon."

I wondered why everything in this area carried a foreboding name. And where did this mongrel we named Dog come from? These questions along with others were placed on the back burner of my mind as we prepared for our decent, carefully checking everyone's backpack to make sure that straps were secure and the heavy loads, which by now had swelled to forty pounds or more, were well balanced. Standing on the edge of the bluff I asked Carlos if he was ready for this, and he replied, "Si, mucho!" With his energetic response ringing in my ears, the eight of us, including Dog, began the dangerous decent into what the locals referred to as the Valley Of Darkness, an ominous name that I hoped would not live up to its image.

For the next two hours we slipped and slid down a narrow, muddy rut that had been cut into the mountainside by running water. This was our trail. Horses couldn't have made it. Hell, we could barely find the traction needed to upright ourselves. With each grueling step our feet submerged into the thick brown, gooey clay. It took extreme effort and every muscle in our legs and backs to step out of the muck. I could feel the blisters forming on the soles and sides of my feet and I would have traded half the gold in

Hell for a hot tub of water and some Epsom salts. The natives were more accustomed to this type of terrain and the muddy conditions, they were always well ahead of me. Slowly, one by one, they vanished from my view, except Carlos who was always watching my back and chuckling as I slid downhill on my backpack to provide some relief for my aching feet.

Carlos took pity in my dilemma and would stop every few minutes and pull out a smoke, giving me a chance to cut some distance between us. He would quickly catch up to me and the ritual would start all over again. Halfway down the treacherous slope, I was carefully navigating a tricky switchback when my concentration was broken by a loud scream that echoed through the canyon. Carlos was in trouble and tumbling head over heels headed straight for me. The only thing separating Carlos from certain death was my tired body. I planted my feet deep in the muddy clay and braced for the impact, while screaming at Ochoa to get out of the way.

The 'parrot' dove for cover about the time that Carlos slammed into my torso feet first, sending me reeling towards the edge of a thousand foot drop. I couldn't find anything solid to grab hold of, just slimy mud and loose sod. Time seemed to revert to slow motion as the two of us tumbled towards eternity and then everything stopped. My feet came to rest on something solid. A large boulder had broken our free fall. Ochoa arrived, calmly surveyed the scene, and then quickly dropped a rope to help us back onto the trail. As we crawled up he asked if we were okay. I could only stare at him. Carlos shook his head and reached for another Full Speed, Ecuador's answer to a Marlboro. We were both out of breath and shaken. My ribs hurt from the force of the blow, but I wasn't about to stop now. Covered in muck, but back on the trail, Carlos stayed only a few feet behind me.

The closer we got to the bottom of the gorge, the steeper the trail became. Footing was treacherous and each of us knew that one missed step could result in death. Finally, one by one we reached the bottom. Everyone was accounted for, even Dog. We took a well deserved smoke break then passed out dried beef jerky to everyone, including Perro who devoured the five-inch stick in one bite. The river with the ominous name had certainly stirred fires within, but the entrance to its treasures was a test of one's spirit and determination. Little Hell made you earn your way and prove yourself with every step.

Adjusting our heavy loads we set out in single file for the Valley of Darkness. Before long we found a small animal trail that led us further and further into the darkening valley. Francisco had cautioned us the night before to be on alert for cats and snakes. Needless to say my eyes searched for sudden movement along the trail. Perro wasn't helping my acute sense of paranoia, darting nervously in and out of the thick brush.

We were halfway to the lagoon. My feet burned with broken blisters that were now seeping body fluids into the strange mix of dirt and sweat, which caused me to wince with every step. I hobbled along in excruciating pain trying to catch up with the rest of the crew. The only one that seemed curious about my snail's pace was Perro, who would stop on the trail time and time again to check my location. He was fast becoming my friend and protector.

We reached Yana Cocha the afternoon of the second day. After pitching camp, Pepe, one of our cargo carriers, gathered dry grass and built a roaring fire for the specialty of the evening: rice, dried pork and white beans. Carlos and I went down to the lagoon to wash our muddy boots when a horrible stench filled my nostrils, "What the hell is that?" I asked Carlos in broken Spanish. "Azufre" he replied. It was the strong and pungent smell of sulfur. Black Lagoon was aptly named. We were camped on the blackened rim of an extinct volcano. The small stream that flowed from the towering mountain nearby fed the small body of stagnant water whose odor of sulfur was stifling.

Something didn't feel right as I splashed my boots in the brackish pool. I stood motionless. It was difficult to overcome an eerie feeling that I was being watched. It was one of those displaced sensations when the mind plays tricks on you when you find yourself in an uncomfortable place or situation. "I don't like this place," I said to anyone listening.

We returned to camp and found Francisco scrutinizing his hand drawn map of the area. He had taken great effort in the elaborate drawing with the strange sounding names. The many rivers that dotted the area were named and traced in blue color and the mountains were also identified and colored in red. He had even marked in dots, a possible helicopter flight path for the future, and at the Rio Infiernilios, our final destination, he had artistically drawn in two palm trees. The bottom of the map was signed Jefe, which designated him as the boss of the expedition. There were x markings he had painstakingly drawn on the white butcher paper to designate our trail. The x markings mysteriously vanished after we passed the Laguna la Respondona, another small lagoon to the east of us.

I asked why the trail vanished at the point on the map that marked the Cordillera De Espadillas; a jagged mountain range the setting sun was beginning to disappear behind. He looked up at me and grinned. The one gold tooth in his mouth glinted in the final rays of the afternoon sun. He seemed proud as he simply said in partial English. "It is no more." "You mean we have to climb that?" I shot back at him. Still grinning, he nodded.

We had brought three two-man tents. I occupied one with all the supplies, food, rope and bateas, which was Ecuador's version of the distinguished gold pan from the days of California's 49ers. Francisco and Carlos occupied the second tent and Pepe, Ochoa, Marko and Octavio fought for space in the remaining tent. Camp was set in a semi-circle with the tents facing each other, which offered the best positioning for security and communication. However this sleeping arrangement had its drawbacks. Just about the time my senses were edging towards sleep, Francisco would begin to snore and as if on cue, was soon joined by a chorus of snores from the other tent. Although dead tired I simply could not sleep. My chest ached, my feet ached, and I felt hemmed in, trapped in the guts of a not so dormant volcano that was stirring.

Yana Cocha was a horribly depressing place. There was no wind. Only silence mixed with the putrid smell of sulfur that burned the nostrils and dulled the senses. Sweating, I tossed and turned until midnight, then, having to relieve myself, I stepped outside into the fog and made my way into some bushes about twenty yards from the tents. The night sky was alive and vibrant with millions of stars. The Valley Of Darkness was being true to its name. Making my way back to camp, I again had the eerie feeling that I was being watched. It was an uneasy feeling that I couldn't shake. I didn't know if I believed in the mountain spirits, but I found myself talking to them, asking them to let us pass in safety. Approaching the camp, Carlos emerged from his tent, "Senor Lee, de miedo, no?" Scary was the right word that summed up the entire evening.

Perro, our wayward and adopted canine friend, also felt uneasy and crouched nearby in the darkness. Our self-appointed security force took his job seriously. He was standoffish, which was probably due to years of neglect and abuse that forged a distrust of the human race. I could understand his caution, then realizing he was cold and scared, I decided to see if I could get him to come closer and eat from my hand which, up to now he had refused to do. He would always sneak into the camp area after we finished eating and grab a mouth full and retreat to safety. He appreciated the food but he wasn't quite sure he could trust these strange men that seemed to be on a journey to nowhere.

I reached for a hand full of beef jerky and positioned my lantern so that he could see my gestures. I extended my hand and in a soothing voice beckoned him toward me. At first, he just sat there some twenty feet away and stared. Then I moved toward him, again extending my hand to show him the food. He moved away. He was still uneasy with my gestures of good will so I returned to the doorway of my tent, slumped down and looked away. Out of the corner of my eye I saw that he was coming closer and becoming more comfortable. Then it dawned on me, I was now an equal as he inched himself towards me. I extended my hand with the food and he finally came in closer. In a flash those fierce looking teeth grabbed the jerky from my hand and once again he retreated into the darkness. Perro brought back memories of my first pet.

My grandfather found a baby coyote wandering aimlessly along the railroad tracks and brought him home as a play pal for me. In the beginning he was snappy, angry, and rebellious; just plain unhappy about being separated from his wild pack. Once he realized that I meant him no harm he settled in and seemed thankful that his every waking moment was not focused on a constant search for food and water. I named him Lobo and raised him from a pup, teaching him to eat from my hand and sleep at the foot of my bed to protect me from the evils of night.

We were inseparable while exploring the uncharted terrain of the desert and gullies surrounding our home in Peridot. Occasionally, we would overstep our boundaries and enter the territorial world claimed by the young Apache braves. With Lobo at my side, I was no longer fair game for these tough young warriors that in the past had made me the target of their hurled rocks. Our battles were real. Before Lobo I was fair game similar to General George Custer at the battle of Little Bighorn, blond, outnumbered, and outgunned. But with Lobo at my side no one dared challenge us. I found this strange until I learned that some of the Apaches said he was my spirit dog and came into my life as my protector. This spirit dog legend gave me newfound confidence, which created a sense of independence and freedom to unabatedly roam our harsh surroundings.

Lobo was the only living thing that really knew and understood me. It didn't matter what my mood or what mischief I was creating; he was always by my side. I could talk to him and spill my guts about my feelings. I knew he would respect me no matter what I said or did. I knew my spirit dog would never let me down and because of this I made certain he had whatever I could offer. This realization was so comforting. He was my stability and security in a world I didn't feel connected to.

Perro had the same mystique and cagey characteristics of my friend from the past. And like Lobo, I knew he would eventually come around once he knew that he could trust me. I made my way back into the tent and tossed and turned for a while before the little sister of death arrived and I drifted into a deep sleep.

With the first gold tones of sunlight, we broke camp and headed for the base of the towering mountain that I knew we would be hard pressed to scale. The upward climb became sheer hell, as we crawled and clawed our way to the top of Little Spades lofty heights. After collapsing on a ridge that took four hours to reach, I had my first look at the jungle that awaited our arrival.

Dropping my backpack to the ground, I tried to relieve the aching red welts and bruises carved into my shoulders by the heavy load. Between my feet, my ribs, and my back I was definitely hurting. Ochoa, who looked like a man that was on the verge of sheer exhaustion, joined me. We were both gasping for oxygen, deeply filling our lungs with the thin air and at the same time coughing up the traces of sulfur residue that still clung to the tissue of our mouths. We sat in stone silence sharing warm fruit for what seemed like the longest time, until, Francisco yelled in his high pitch voice, "Vamos!" It was time to move on.

We strapped on our burdensome packs, took a deep breath and sucked it up once again. Although we were exhausted, the Little Spades mountain ridge was a godsend. We were away from the valley of depression below us and on a semi-jungle trail that was somewhat level and easy to navigate. My feet were dry and the salve that Francisco had applied to my blisters was helping them and allowed me to stay close to the pack of fast moving Canari.

Ochoa was on my heels when Francisco called for our band to halt. We set up camp in a small clearing on the ridge that was designated on our map as a helicopter-landing site. Carlos was tired but jovial, "Only one day more," he chirped. We gazed from our lofty perch into the sea of green that stretched out before us under the magnificent rays of the setting sun. "La selva," Carlos said, while pointing in the direction that our eyes had been searching. "Oriente," Ochoa echoed from the distance. Then the two men got into an argument over the proper term being given to the endless green growth beneath our makeshift camp. In my mind it was still the jungle and tomorrow we would get our first taste of it. Ochoa pointed east. "Senor Lee, you no go there. In Oriente people called Jivaro live. They take heads. They capture people's spirit. Mucho, mucho tigres and snakes there. Oriente es muy peligroso. Don't go there, Senor Lee."

I was beginning to understand why so much of this country was untouched and still virgin. If the natives were too fearful to venture into an adjoining providence then I understood why the colonials in the cities weren't avid explorers. Ochoa's concern was real. "Don't worry Ochoa; I have no reason to go there. All the gold is here...in the mountains. We'll be rich men in a few days."

Even the able bodied men of Nabon did not want to know about their neighbors who in year's past had been head hunters. And if they did, they would have problems with the terrain. Getting anywhere in this country was a problem. Indians who were in much better physical condition than my self were having a tough time. I couldn't imagine trying to make this journey without tents, weapons, backpacks, or ample food supplies to guarantee staying power. It was too much work to only stay a day or two. I was beginning to understand why only a few had ever made this journey.

The boys were in great spirits as they regained their energy and collected their thoughts. Occasionally they would speak in soft tones but for the most part they too were lost in their thoughts of a life's dream that was slowly nearing a tangible reality. They would enter Hell tomorrow and possibly for the first time, they had achieved purpose and dignity.

Dinner consisted of rice, beans and mutton and was washed down by hot, steaming Ecuadorian coffee. The boys were in a good frame of mind, laughing and joking with each other. Even Perro, who had been nervous the night before was in a good mood while waiting his turn for left over scraps. We were all glad to be out of the valley of darkness. The slight breeze rustling through the tall bushes filled our nostrils with the freshness of the pure Andean air. It felt good up here under the clear skies that would soon transpose themselves into the night magic of the universe. At Yana Cocha time seemed to stand still.

It was difficult to believe that a short time ago I was embedded in a puritan society complete with suit and tie while arranging million dollar loans for commercial developers in their push for progress. Now I was in the company of men that would never comprehend what I left behind in my search for adventure. If they had understood I'm sure I would have immediately been labeled as poco loco—a little crazy. Their whole life was a struggle to find security for themselves and their loved ones. Just to know that I had given up the elusive security they craved, to join them in an elusive quest for El Dorado, would have left them shaking their heads and suggesting that I see a doctor.

My friends and family had already cast their vote and wondered what was wrong with me. And even though I was more comfortable in this place, at this time, with these men than I had been in any social circle, I had to admit that it was strange to be encamped on the fringe of a jungle with the direct descendants of a once flesh eating tribe.

The ancient Canari ruled with authority and with purpose under their great leader, Duma. They occupied the highlands and coastal areas. At its peak, the Canari culture extended as far north as Nudo de Azuay and south to Azogues. Chroniclers recorded that they resisted the Inca incursion for over 100 years in hostile, open warfare that decimated whole armies in a bloodbath that sapped both empires of their most precious commodity: young warriors. Unable to subdue the Canari, the Inca negotiated a political settlement with the fierce tribe and then years later the Conquistadores came and did what their name implied; they conquered. The rest is history. Unlike the Jivaro tribes to the east, the Canari attempted to blend into the European influence and instead were enslaved in a horrific caste system.

The next morning we struck camp early and for three hours hacked through a jungle of semi-darkness. Reaching the end of the mountain ridge we began our plunge into a maze of even thicker vegetation. We found ourselves surrounded by green jungle that the boys referred to as el tubo. The Tube got its name because we were literally in a tunnel chopped and hacked from the densest growth I had ever seen. Above us a thick jungle canopy of various shades of green blocked out the sun and sky. Indians in front with razor sharp machetes hacked a small opening wide enough for men and backpacks to slither through. The Canari had told me the truth when they said the trail was over grown from lack of use. The cool morning breeze that filtered through the leaves made our tough decent down the mountain slope bearable. But, after a couple hours of constant stooping under the fresh cut vines, I wished that I was either shorter or the guys up front that were chopping were taller.

The last limb sprang into place with a hard snap behind us, and we stepped cautiously into a small clearing in front of a slow moving river with a sandy bottom. Finding a smooth place, I dropped my gear and checked my watch. It was noon. Ochoa joined me and we both lit up while nervously eyeing the thick jungle across the river on an adjacent ridge. Perro crouched behind me, alert, watching my back. We were both dreading another climb, when Carlos came running up to us, with a big smile on his face. He pivoted on one heel and waved enthusiastically towards the peaceful water. Shouting in broken English, "Senor Lee, welcome to your Rio Infiernilios!" We had finally arrived.

We waded across the waist deep river and selected a knoll for our permanent campsite that offered elevation above the Rio and an ideal view of over seventy feet of riverbank. While two of our men started cutting back the jungle growth to fit our tents and cooking area, Carlos guided me upstream about twenty yards where we came to several slow moving bends in the river. "Bateas aqui!" he demanded. I combed the shallow water for a glint of the precious metal, when sudden movement in the river startled me.

Carlos reacted on instinct and with one swing of his machete; he decapitated a large brown fish. "Trucha," he yelled, proudly displaying his catch. The river had trout. Now, if it only had gold everyone's dreams would come true.

It was amazing what the lure of El Dorado can do for tired and worn out bodies that ached with every motion. Not wanting to waste any daylight we barely controlled our excitement and decided to go to work immediately. The three washers unstrung the large bateas lashed to their backpacks while Marco and Pepe began clearing the campsite of underbrush and Octavio gathered firewood. It was exhilarating for all, especially for Carlos and Francisco who had dreamed of this moment.

Out of enthusiasm, I offered a one hundred sucres bonus to the first man that found a gold nugget larger than a match head. The three gold washers let out yelps of joy that startled Perro who was standing on the riverbank. They began jockeying for position along the beach. I located the tweezers and empty coffee cans and stood by, waiting for the show to begin. Watching the eager men unfurl their wooden bateas, I made a mental bet with myself as to who would find color first. Ochoa was already in the cold water with his batea. Then my eyes went to Francisco, who was now in the river and washing between two medium sized rocks. I decided Carlos would win hands down. Why Carlos? It was simple logic. He prided himself in being the best.

I was right! Carlos let out a yell, which resembled, "Ajiie!" Jumping into the water I waded over to Carlos and at the bottom of his wooden batea was a gold nugget about the size and contour of a flat and smooth contact lens. It had traveled a long way from its original source and was polished by the tumbling, churning waters of the Infiernilios River. Carlos had found it in a small eddy next to a large rock, which had stopped its long journey; at the very spot he had shown me earlier. I told him to hold that pose, scrambled up the bank, grabbed the camera from my backpack, and returned to his side where I shot my first photo of gold. Carlos beamed while Ochoa and Francisco pouted, so I filmed them too, along with Perro in the background taking a leisurely swim.

For the next three hours I anxiously witnessed an amazing primitive art form. The large wooden bateas came alive with their own magical flow of movement. Dipping their bateas into the slow moving water along the riverbank the masters of this art form equally divided water with sand and mud. First it swung in a clockwise motion and then counterclockwise. They swirled the water in the bateas until the small rocks, mud, gravel and other debris had been separated from the heavier contents, in the bottom of the wooden pan. Gold, the heavier element worked its way to the bottom and lodged in a small circular groove carved in the base, which was the trademark of the batea. Once the gold had settled in, it would be extracted. If the gold was in tiny granules, similar to what we found in the Rio Shingata, then it was removed using a magnet to separate magnetite from the ore. If it were larger, tweezers removed it. At Little Hell the nuggets were removed from the bateas with your fingers. They had described the gold in Infiernilios as being pepitas, in fact that was the only word they had ever used. I understood why. The river was abundant in gold nuggets of various shapes and sizes.

Although Carlos won the contest and pocketed one hundred sucres for his efforts, Ochoa and Francisco were not far behind. Ochoa had moved his location to a sandbar and was hitting pay dirt. In a relatively short period of time he pulled out over a dozen glittering pebble size nuggets. Francisco had the record for size and weight having found a coarse nugget that we estimated to be roughly three ounces. The texture of the nugget told us that its source was close by, perhaps a vein in the bank of the Infiernilios. Immediately I placed Marko and Pepe on the side of the banks with picks and shovels to break away the rocky ledges that framed the river on a search for veins of gold that possibly fed into the river during high water. There were many other nuggets smooth to the touch that had most likely tumbled from a great distance. The old adage, smoother is farther and courser is closer was a philosophy that originated in the old mining camps of yesteryear and it held true today.

My champion gold washers soon began to tire. Their lips and skin taking on a bluish hue from the hours spent in the icy cold water. It was time to relieve them. Marko, Pepe and I grabbed bateas and waded in. In the beginning my lack of rhythm caused me to slosh rather than swirl the heavy wooden batea. Carlos saw my dilemma, moved up the riverbank and began shouting instructions while urging me on. "Senor Lee, more sand and less water." The trick was a three-quarters mixture of sand or gravel and one-quarter water. The art of the clockwise and counter clockwise spin that settled the heavy gold into the bottom of the batea was the final act that insured success.

The boys seemed to be getting a kick out of my fumbling efforts and after a few private jokes and some snickering they were amazed to see how quickly I picked this art up. Before long my efforts were paying off and when I found my first nugget my excitement was unchecked; I was like a kid on Christmas Eve waiting for Saint Nick to arrive. My Santa was a course gold nugget about the size of my little fingernail. "Hey, this is a piece of cake," I thought, until after twenty or thirty minutes and my body began rebelling against the cold water. My lower body and hands started to experience numbness. Then my mind began to rebel and the only thing I could think about was getting warm and dry. After a while the mind must send a hidden command to the body and tells it to 'Cool it; my host is dumb or stubborn or is just a glutton for punishment. So body you're on your own.' Somehow the body understands all of this and adapts accordingly. After a while I stopped thinking about the cold water and began concentrating on removing the tears of the sun from a giving river. The mountain spirits had accepted us.

Perro was busy too, barking and lunging at the brown mountain trout in the river upstream. Francisco remarked, "Perro tired of rice and beans too." As the day wore on, Carlos still held the consistency record in quantity and quality for nuggets, or what the Indians referred to as pepitas. By late afternoon, the rays of the sun were becoming obscured by the shadows being cast by the Cordillera Manga Hurco. It was getting colder and a slight breeze added to the discomfort of the wet clothes clinging to our aching bodies. But, for all the physical displeasure we were still a happy bunch of campers. Now I knew how the '49ers hung in day after day because once you have really experienced gold fever it's as addictive as any drug. And once you have hit a "glory hole", as the old-timers called a major source of gold usually found in a gold bearing river, then you were hooked for life. Before night set in our glory hole had filled two small coffee cans of precious metal.

It was true. Little Hell was an abundant river that had never been worked by the Inca or the Spanish. Carlos said I was the first white man to set foot on its soil. I believed him. I also believed the story they told me about one Canari who took out six hundred grams of gold in two days. If one gram equals 1/28 of an ounce that meant that he washed out almost twenty-one and one-half ounces of the precious metal. He was probably paid less then three hundred dollars for his backbreaking work. But, in this country where the daily payment for an Indian was a dollar a day the man made almost a year's salary in two days. I understood why this area had been such a closely guarded secret.

Carlos said one gold nugget found here weighed over one-hundred-and-eighty grams or roughly six and one-half ounces. They also spoke of three other rivers nearby, which were abundant in the precious metal. They were the Pepitas (nuggets), the Espadillas (Spades), and the Manuolco (no translation). They said that the Manuolco rivaled the Infiernilios in gold, but it was very difficult to explore because of the dense trail-less jungle that it flowed through. We decided to pass on that one, but if time permitted we would sample the Nuggets and Spades River. I asked why this river was named Little Hell and was a bit uncomfortable when they could only offer a weak smile.

After dinner we started taking inventory of our remaining food supplies. We estimated that if we could eliminate breakfast for two days and cut back on lunch, and add fish to our diet, then we could squeeze in another two days on the river. We wanted all the time we could muster on Infiernilios. In four hours we had managed to fill two coffee cans of gold nuggets. Were we riding a lucky streak or was the river this prolific? I opted for the ladder and hoped we would be able to retain a consistent output during our limited time at Infiernilios.

I knew that we would be limited to an eight-hour workday because we had no wet suits to protect against the biting cold of the river and the fact that evening came early because of the dark shadows from the mountain range to the west. I estimated that two extra days on the river should equal about 16 hours of gold washing time, which when multiplied by a pound of precious metal per hour equaled a potential take of 16 pounds of gold. The gold was jewelry quality nuggets, which would bring a higher rate than $35.00 per ounce if sold to the right person. After subtracting the thirty percent promised to my jefes, Carlos and Francisco, and the additional ten percent to the other four men, we could all have a fabulous take for three days work.

Under the dim light of my battered lantern Perro finished devouring the left over trout and rice while the Canari, tired and spent, slumped in their tents seeking much-needed rest. I hunched over my notebook scribbling incredible, yet factual calculations, of what a four-inch suction dredge would be capable of recovering per day on this plentiful river. "12 cubic yards per hour," I kept mumbling to myself. "My God," I said to Perro who had become my tent mate. "We've got to get a helicopter, amigo." With that final comment I crawled into my sleeping bag and blew out the lamplight as incredible dredge statistics sailed through my mind. It was a highly emotional day for everyone and I couldn't wait for tomorrow to have another go at the river now affectionately being referred to as Little Heaven.

It was shortly after midnight, when I was awakened by thunder and the odd, pit-pit-pit, sounds of large raindrops, slapping against the tent. Not giving it much though, I drifted back into a deep sleep. Just before dawn I began to experience a dream that was so real, I thought I was going to die of suffocation. My breathing was shallow and laborious. I gasped for air as I felt heaviness on my chest and face. It felt as if I had been buried alive in a wet, cold sarcophagus of strange material that was sucking the life out of me. Struggling through sleep to gain control of my fate, I started flailing my arms and legs and then popped into a semi-state of consciousness and reluctantly opened my eyes. To my dismay, I discovered that my nightmare was real. Something heavy and wet was on top of me, slowly squeezing the breath from my lungs.

My arms pushed against my attacker as I tried to free myself. Using my right arm with all the strength I could muster, I lifted the weight from my chest and started sliding my body to my left, towards Perro who was sound asleep and snoring in crescendo with the deafening thunder. "Perro", I yelled. He didn't hear me over the storm. Finally, with my remaining strength, I lunged and managed to slap him on the head with my free left hand. He bolted awake, growled, then barked, and then he seemed to realize what was going on and started whining as if to provide moral support while I wiggled a path to freedom. In the early morning light I saw that the right side of my tent had collapsed from the weight of the torrential rain.

I had renamed the river Little Heaven, but she showed her other face; that of an ugly, unruly devil. Tons of water cascaded in a mad rush down stream. The small arroyo was filled with a deafening noise that reminded me of the freight trains that jarred me out of bed as a youngster. Fortunately our camp had been placed on high ground, but, unfortunately, placed on the wrong side of the turbulent water. As my team stood on the knoll watching Infiernilios live up to its name, I had a gnawing feeling that we were into something we couldn't handle. During the night the river had doubled in width and depth causing our escape route to vanish in the turbulent brown water. The mountain spirits had changed their mind; we weren't welcome.

Sheets of cold rain peppered our faces and blurred our vision. We managed to take a head count and salvage what we could.

Everyone was accounted for except Perro, who was nowhere to be found. Marco said that he high tailed and when last seen he was heading into the jungle. I barked orders and gathered our gear knowing that my statement was as much bravado for the men as it was for me. On the surface I was semi-confident that we would make it out, but I had my doubts.

The first day we were marooned we cursed our luck. The second day it was the river. On the third day we cursed the heavens. Then things began to get serious. Our food supplies were dwindling and without dry wood to build a fire we were down to one meal a day, which was hard jerky. The good news was that Perro had returned to camp the day after he left. The bad news was we had another mouth to feed. But everyone chipped in and our camp mascot got his rations, too. Our days and nights were spent huddling in the water soaked, musty smelling canvas tents. Everything was wet... food, tents, shoes, socks and our clothes, if you wanted to call them that. They clung to us, cold, muddy and wet and they smelled of mildew.

We hadn't seen sun in three days and that was beginning to wear on us. Our machismo confidence was starting to waiver as it rained constantly from downpour to shower and then to deluge. During the periods of light rain we searched up and down stream for a natural bridge. None was found. On the fourth day, the rain had let up somewhat and out of desperation we tried a dumb move. We let Francisco talk us into trying to swim to the opposite riverbank with a rope tied to his side. The moment he waded into the swift river; he was swept off his feet and carried down river until the rope stopped his progress. He almost drowned in the churning water before we could reel him in. Coughing and sputtering, Francisco began to weep as he made the sign of the cross on his bare chest.

On the fifth and sixth day the sun shone brightly in the equatorial sky, yet Infiernilios rolled on with its own brand of defiance. We, along with our sleeping bags and tents were drying out but we were out of all staple foods, including rice, beans, yucca and crackers. Only five bags of freeze-dried applesauce remained along with six sticks of beef jerky. Our fresh water supply was down to one canteen and two cooking pots full of rainwater. Our strength and stamina had been sapped through worry and lack of nourishment. To a man, we would have traded all the gold in Infiernilios for our freedom.

The human is a predictable creature. When in trouble humans question the material versus spiritual values, but not before. I questioned both. I also questioned my lack of exit strategy. I had been relying on the Indians to come up with the answers, after all this was their territory. They in turn were waiting for me to solve the problem. We were in a state of reaction waiting for each other to come up with a solution. In the meantime the river was dictating our fate. Realizing this, I called for a meeting and we began a plan of action. "Francisco, how many days before the river will return to normal?" I asked.

Looking first to Carlos, who shrugged his shoulders, then to the others that shook their heads? "Quien sabe?" he said with a shrug.

"Who knows? Ok, if we don't have a clue as to how long we might have to wait, then the waiting game is over. Let's get the hell out of here," I said defiantly. Everyone agreed. We laid the map on the dry tarp and began examining the terrain, beginning with our present location. We were stranded on the east side of the Rio Infiernilios and the village of Nabon was due west. Since we couldn't cross the river and head west we were left with only two options. We could head downstream towards the confluence of the Infiernilios and the Amarillo, which would probably be an even more impossible scenario. The runoff there would be worse than where we were. Our only choice left was to head upstream, hoping that the river would become narrower and smaller as we approached its source somewhere in the rugged Cordillera Manga Hurco.

We planned our escape to coincide with first light. Knowing our search for an opening across the river would require all the remaining energy we had, we would retire early and try to get a full nights rest. Inside my dark, foul smelling tent, it was obvious from the muffled conversations coming from the other tents that the boys, like me, couldn't sleep. My mind raced from one dilemma that we were facing to another as I tossed and turned. I wondered where Perro had gone. Would he return in time to join us in our attempted escape or would he be left behind?

Childhood memories again comforted my mind. I thought about Lobo and our adventures. It always felt good to be out in the open spaces even if I was on the alert for rattlesnakes that lived on the San Carlos Apache Reservation. I was learning the way of the Apache, making bows and arrows and speaking a few words in their native tongue. My exploration took me to the lofty heights of a mountain ridge where I found a teepee far removed from other Apache dwellings. Unusual markings and symbols adorned its doorway. Always curious, but also cautious, I edged my way closer trying to remain as stealthy as my Apache brothers had been when we hunted for quail and rabbits. Suddenly an old Apache man appeared from behind a large bush. Angry and outraged by my intrusion, he confronted me. His towering figure frightened me out of my wits. I backed off searching for an escape route.

When the old brave realized that he had taken me to the threshold of fear, he loudly let out a wild burst of uncontrollable laughter and began beckoning me to come forward. He smiled and, even though my heart was pounding like the rhythm of those distant Apache drums that kept us awake on Saturday nights, I finally began to relax. He told me his given name, which I could never remember or pronounce, so he asked me to call him the Apache Kid. I asked if he was the Apache Kid? He answered with a smile.

Decades earlier, I might have been just another trophy hanging from his scalp belt. But, on this day we were just two lonely people sitting cross-legged in the sand. He held a crooked stick in his hand drawing circles and crosses, trying to speak in a mixture of English, Spanish and his native tongue. I learned that he was in his eighties, as best he could recall because birth records in those days were non-existent for those of his kind. I asked again. "Are you the real Apache Kid?"

More circles and crosses and then he said sadly that those days were gone forever. Then, with head hung low, he became silent. He looked deep into my soul with eyes that reflected the all-knowing experiences of a man that had lived life to its fullest... At least that was what I imagined. We must have talked for hours.

In a mixture of three languages he explained that he had paid for his errant ways and the crimes against the bluecoats. Now he was just an old warrior with nothing left except fading memories from his past. He seemed to know his days were numbered and the destined time to meet his Creator was drawing near. Before long it was time to go, but I promised him I would return.

It was exhilarating to keep my newfound friend a secret and with every opportunity, I would sneak up the hill and spend hours with my mentor. The Kid taught me the Indian ways of survival. He showed me where water could be found in the parched desert by simply studying nature. A barrel cactus was a disguised pulpy canteen for the precious fluid. He also taught me how to recognize the desert denizens by the tracks they left in soft sand, including the rope-like imprints indicating a snake had slithered by sometime earlier. The rattlesnake imprint was unlike the sidewinder and they were different from the Gila monster. He instructed me how to select a rock with balance and the correct size to be used as the blunt end of a tomahawk when tied to a strong stick with leather ties.

Most kids my age had their own set of idols that were usually movie stars or sports figures or both. Mine was a small bow-legged man that accepted me as an equal rather than another white-eye that was petrified of the "savages" that lived and roamed on the reservation. When he spoke, I was often forced to lean forward because of the softness of his voice. He spoke of his youth, his dark eyes growing moist and twinkling with excitement. When he spoke of his warrior days, he would swell with pride, but the eye twinkle transformed into anger. The rage subsided into sadness when he gazed across the barren landscape known as the Rez. "The blue coats took away our land and gave us only what they did not want or need. They took away our language. They took away our dignity. It's gone. But they can't capture our spirit," he said proudly.

Then he asked me a startling question. "Where is your spirit dog?" Before I could answer, he began a long story that I took as part real and part mystical about Lobo's heritage. My eyes widened in expectation. He began by describing his second wife, who was from another tribe and a member of the Coyote clan. The coyote was revered as a powerful spirit. Cunning and brave it only appeared to those in need of friendship and protection.

"Many Apaches believe that the wily coyote is only a prankster in search of mischief and mirth and not to be fully trusted, but for me this was not true. Because as a young boy, I befriended a young pup as you did and raised him. Later on when I went to war with the bluecoats he accompanied me and looked out for me and on two occasions he saved my life and then gave up his own so that I might live. Once when the soldiers came close to my camp, he sensed they were there before I was even aware of their presence. He drew them away from my camp by barking and running into the desert and they took out after him with their rifles and shot him for sport. Then they laughed and congratulated each other on their marksmanship and false manhood. I lived and he died. But, my little brother, he only died in body only because his spirit still lives with me." Then he pointed to a clay bowl in the corner of his teepee that contained dried scraps of food.

Noticing the tears welling in the corners of my eyes he reached over and gently patted me on the shoulder, while at the same time speaking softly in broken English. "Don't ever worry my little friend. Your spirit dog will always be with you and he will rise to the occasion in your time of need. My spirit dog comes more often to me now and tells me that my time is drawing near. But, I am not sad, only happy that we will soon be together to explore the great world. We will be free to come and go as we wish without any interference. And no human will be playing Creator with our lives and our destiny." Then he grew quiet, lost in his own thoughts.

Lobo and the Kid had shaped my life and taught me to survive against the odds. I knew it was time to employ the lessons of my youth. Staggering out of our tents at first light, we placed Francisco's makeshift map on the soggy tarp. The stakes couldn't have been higher and yet, we might as well been reading tea leaves. None in our crew had been up the Infiernilios to the headwaters and what awaited us was anyone's guess. Just staring at Manga Hurco was a will breaker. We would have to travel light, as we'd be climbing through muddy terrain. Tents, which were in bad shape anyway, would be left behind with the shovels, picks, and other tools. Carlos tossed his tattered and empty backpack and I tossed my camera, which had been damaged by the rain. I wrapped the one roll of film that was shot the first day in a semi-dry sock and placed it with the map and the coffee cans containing the gold, in the remaining backpack. I checked my watch; it was almost 10:00 a.m. It was time to move out. I knew that with a little luck, we might reach the base of the mountain by nightfall.

Just as we geared up, we heard the faint sound of a dog barking in the distance. It was Perro and he was trying to get our attention. We made our way downstream. The barking became louder and more animated. As we rounded a small bend in the river, there was Perro, standing on a small knoll, barking ferociously, and staring at something in the river. We looked in disbelief. There was a large tree trunk wedged between two massive rocks that had formed a makeshift bridge. I felt the burning of salty tears in my eyes as I ran over to Perro and gave him a much-deserved hug. The boys were yelling and screaming at the top of their lungs, and waving their machetes celebrating our good luck. Ochoa had no love of animals but his eyes were as big as saucers when he asked, "Senor Lee, how he know?" He petted our dog-named Dog.

"Maybe he's a spirit dog?" I replied in a voice breaking with emotion.

Grabbing a rope and tying it around my waist, I stripped off my boots and shirt and moved cautiously towards the slimy, algae covered cedar log. I yelled for more slack on the rope that was being anchored by those that still had strength. I had to wade into the strong current to a large boulder, then somehow hoist myself on top of the wedged tree trunk. Surprisingly, I found the going much easier than expected. I reached the anchor boulder, but accidentally stumbled on another boulder hidden from view by the broiling current. Using that rock as a stepping-stone I was able to straddle the trunk, which was solidly wedged in place. My compadres gave me the slack I needed in the tether line and inched slowly across the river to freedom.

I secured the rope and anchored it to a thick tree above the playa and parallel to the tree trunk. Carlos and Francisco did likewise on the other side. Now we had a makeshift rope bridge to hold on to above the tree trunk bridge. It helped the team balance and gave them a lifeline as they crossed. However, there was a major problem. How would we get Perro across? He might be able to cross the log on his own, but if he slipped and fell into the raging water he would be swept downstream to a watery death. We needed someone to carry him to safety, which meant that the person carrying him could only use one hand on the rope and the other hand wrapped around Perro's waist. And to make matters even worse, he was not a small dog and he might freak at being suspended in space.

As I was preparing to cross the river to bring Dog over when Ochoa, the animal hater, grabbed Perro around the waist and with Carlos helping, made it to our makeshift bridge. With one hand as support, he and the docile K-9 safely crossed over. I reached for them as they made it to the escape side of the river. Grabbing my hand Ochoa, with a hint of reverence, said, "If he spirit dog, I want be close at him on damn bridge,"

One by one, my scarecrow looking crew made their way across the Infiernilios, not even pausing to look back and instead pushing into the tubo of the mountain rainforest. We slashed and hacked our way through dense jungle sometimes crawling on our hands and knees through wet and decayed foliage until we reached the summit of Cordillera Espadillas. We paused on the trail to get our bearings, I gazed at the river called Little Hell and couldn't help thinking about how lucky we were to make it out. I could feel my heart skip a beat and thought, "The Kid and our spirit dogs were looking after us."

I left the crew in Nabon and returned to Guayaquil to replenish my strength and recover from the physical and mental scars brought on by the traumatic ordeal at Hell. But, first it was time to sell our poke to a local jeweler. I was certain the storeowner would want to negotiate the price and was pleasantly surprised when he hastily paid my asking price and retreated to the back room. I hadn't checked into the hotel yet and was covered in dried mud and sweat. I thought, "Maybe that's the way to do business in this place? Fourteen days without a hot bath will get you whatever you want." It didn't hurt my cause that he was buying some of the most beautiful gold nuggets ever seen.

When you sold gold the asking price varied depending on whether it was dust, granules, small nuggets or large nuggets that were being offered. It was the look in the jeweler's eyes that betrayed his power of negotiation. The jeweler had never seen pepitas this size with the smoothness and texture. Before I left the jeweler asked if I would be returning with more of the same in the near future. I smiled and casually nodded an affirmative. I wished him well and headed towards the nearest bank to deposit a backpack full of those unpredictable Ecuadorian sucres.

I went to the telegraph office and sent an urgent wire to Nabon. Our code was simple, designed to defeat prying eyes and larcenous hearts and yet give comfort to those that waited for their shares from a white man they barely knew but trusted. It simply read as follows: I have succeeded in selling the property. Stop. I will see you soon. Stop. Rubio.

I smiled with a sense of accomplishment, knowing that there would be a celebration in the town of Nabon. Trust is a double edge sword and we had forged its true meaning during troubling times on the banks of Little Hell. Now that I had completed the circle and fulfilled my obligation, I hailed a cab. It was time to once again become civilized, which included warm meals and a hot bath with real soap.

CHAPTER 2
BETRAYAL

Lee Elders and Bob Olson with friends in Nabon

Every explorer needs a partner. There was Lewis and Clark and Stanley and Livingston. There were also many others who blazed new trails with someone they could trust and depend upon when an untimely situation dictated a friend in need. I needed someone who knew equipment construction and operation, and most important, a person I could trust, with a sense of coolness under pressure and humor combined with the will to succeed and survive.

One such man came to mind, which prompted me to place a phone call to Yuma, Arizona. Bob Olson was an appealing throwback to an age when a man's word was his bond. I wasn't surprised when he came running. At forty-four, he was as strong as an ox and wise to the ways of the world. Prematurely gray with piercing blue eyes, Bob had a sunny personality that could charm the balls off of a pool table and a mind that absorbed everything. I met him when

I was a commercial loan officer for Western American Mortgage Company and he was one of our most revered customers, obtaining millions of dollars in mortgage loans for his bustling custom home business in and around the Yuma Country Club.

Bob was an ex-boxer with a crooked nose that verified his years in the ring. He was also a custom cabinetmaker, construction worker and pilot. The fun loving Swede loved flying his Beech Bonanza, was an adventurer and gambler at heart, and had parlayed a small custom home business into one of the largest in the state of Arizona. Once he had been asked to define pressure; he thought for a few seconds and then quoted one of his idols. "Lee Trevino was once asked the same question and he replied that pressure is having a twenty-foot-down-hole-put, with five dollars in your pocket on a hundred-dollar bet." That was Bob at his best. He loved taking chances and searching for new adventures.

Bob arrived in Guayaquil resembling a golfer that took the wrong turn. His colorful shirt and contagious smile lit up the airport, but he was all business. "When do we begin churning up that river to get its gold? I need some adventure in my life," he said laughing. He certainly was in the right place for a taste of raw adventure. The combination of freedom from a stifling business environment and the lure of El Dorado were the ingredients that prompted his decision to join me. Also, despite the age difference we had developed a strong friendship, almost like brothers.

In between calls to room service and different equipment suppliers, Bob took charge and lined up engines and foot valves for the new dredge while I slept for almost two days. Finally, when I began to return to the living, we decided to venture out for something called a meal. I was looking forward to sitting at a table, on a chair, with real silverware and linen napkins and someone providing a menu with a varied choice of food. We agreed no rice and no beans as that would be our staple for the duration of the expedition.

Hailing a cab, I asked the driver to take us to the finest restaurant in town. Soon we were at the Guayas Bay Club, a private nightclub, gambling casino, and restaurant. After a hair-raising ride the cab driver accompanied us to the front door, tapped on it, and soon a face appeared in the barred peephole. A few quick words were exchanged and we were allowed to enter. The first floor consisted of a very dark cocktail lounge and bar, a perfect place to have a secret rendezvous with a mistress, girl friend, or someone's wife.

The gatekeeper, a hulking, big man in his forties, escorted us through the dark area and up a flight of stairs that were lit well enough to avoid accidents and lawsuits. The top of the stairs opened to a spacious, elegant dining room but before we had the opportunity to take it all in, we were greeted by a stylish, attractive woman in her early fifties. She introduced herself as Barbara Marsh, a transplanted American from New York and the manager of the club.

When Barbara discovered that we were from Arizona and about to leave on an expedition into the wilds of the Andes, she went out of her way to make us comfortable. We knew this would be our nightly hangout when we were served an incredible platter of giant sea prawn the first night. The second night was lobster and so on. Barbara became a good friend and shared parts of her mysterious life with us, while educating us on the politics and machinations of this strange land.

Guayaquil was a sprawling city of intrigue and drama in the 60s. Once while we were downtown searching for a foot valve for the dredge, we had to take refuge in a department store to escape the clouds of tear gas that filled the streets. The scene was right out of a Max Sennet silent film! A flat bed truck rolled up, carrying helmeted, khaki clad, riot police with nightsticks and tear gas canisters. They had been summoned to put down a leftist street demonstration that was anti-U.S., anti-Vietnam. The cops sprayed everything in site, including bystanders. Then to their surprise, the riot police discovered they had forgotten to bring their gas masks. The leftist students realized that the battleground was equal and mounted a charge against the coughing, sneezing riot police. A donnybrook ensued. This city was good for a million laughs or a million tears, or both at the same time.

Guayaquil, a bustling port city straddled the Guayas River and was called the Pearl of the Pacific. Its name was derived from a renegade Indian chief, Guayas, who had refused to surrender to the Spanish Conquistadores. The great chief refused to admit defeat and in an act of self-defiance, he killed his wife, children, and himself. Legend said that before he died he placed a curse on the land. I always wondered if there was some truth in that story because the city has thrived on misery, death and destruction for decades. The Pearl of the Pacific had a history of fires, earthquakes, pirates, and battles between warring generals and political opponents. In 1907, the legendary explorer, Colonel P.H. Fawcett sailed up the Guayas River and in his diary, he noted, "That Guayaquil was a veritable pest house."

It hadn't changed much when it came to the pest problem. The city reeked of an abundance of genus rattus and my friend, Bob hated rats. When the sun set the rodents came out to play and search for food in restaurants, on the streets and in the theaters. Unfortunately the only entertainment after dark was the local movie houses. We were sitting in one of Guayaquil's finest theaters watching a melodrama titled "the Mephisto Waltz" which was in English with Spanish subtitles when Bob let out a blood-curdling yell that not only caused the house lights to go on, but damn near gave me and those seated around us heart attacks. A large rat, about the size of an undernourished cat had crawled up his pants leg after a candy bar he was holding. From then on anytime we went to a picture show, the Swede propped his feet up on the seat in front, whether it was empty or not.

In between good food, bad movies and avoiding the pesky vermin we finally located all the necessary parts for the dredge, including the large truck inner tubes that would float it. I made inquiries about the availability of helicopters and was not shocked to find out that there were none available. Texaco and Shell had whatever was available tied up on long-term leases. Truthfully, I was somewhat relieved to learn this; I knew it was imperative to stay under the radar and low profile if we wanted a fair chance at the riches of Little Hell.

Pleased with what we had accomplished I sent a cryptic telegram to my crew in Nabon that the river raft had finally been built and Uncle Roberto and I would be arriving for a family reunion next week, and that we would be bringing gifts for the wedding. This told them that I had sold my poke, the dredge was constructed and I had a new partner. It also translated to the trusting crew they would receive their shares in the coming weeks.

To celebrate the completion of phase one of our project we left the hotel at dusk and unable to find a cab, we decided to take a shortcut and walk the four blocks to the Bay Club for our final gourmet style meal. Within minutes we found ourselves in an area that was littered with male bodies. Some were sitting staring at us, while others were curled up asleep on makeshift cardboard beds. We shook our heads in dismay at the plight of the poor dregs of this society and their sordid living conditions. We wondered how they could survive an evening with the rats, roaches and what ever else that appeared after dark. Needless to say, our appetites had diminished by the time we reached the club.

Barbara greeted us with her familiar, "How you doing boys." We were not our normal boisterous selves and she asked if there was anything wrong. We told her about our evening stroll and what we had witnessed along the back end of Malecon Boulevard.

"Oh, my God," she cried, "You didn't walk through there after dark, did you?" We assured her that we were fine, only curious and saddened by what we had stumbled into.

"You were in one of the most dangerous places in this country. Those men are stevedores, cargo loaders for the banana ships. They make a dollar a day, for backbreaking work that makes them old men before their time. That's if they live that long," she continued. "Many die early from living on the filthy streets; others die from snakebite."

"Snakebite?" Bob and I asked at the same time.

"Yeah, there's a small green snake, deadly poisonous. It lives on the banana plantations. During harvest the snake is brought in accidentally with the bananas. These poor cargo loaders are bitten on the neck, on the back, even in the face as they carry these huge stalks up the gangplanks and into the cargo hold of ships." Bob and I started squirming in our seats. "Don't go back there," she warned with a smile and a shake of her head. "Those guys are desperate. Most of them are Indio's brought in from villages to work for the land barons. They have nothing to live for. They'd cut your throat for a dollar, if they thought they could get away with it."

The next morning we flew out of the oppressive city. We couldn't wait to escape the sweltering heat, the pestilence, the street beggars and the poverty that made up this teaming city. We circled for altitude in the Saeta turbo prop and I knew that I was ready to trade the small comforts of civilization for another go at an unpredictable spirit shrouded river.

The flight was forty minutes in when it became "white knuckle" time as the experienced pilots jockeyed the plane between two mountain ridges to set up their final approach for a landing at the Cuenca airport. We flew in at treetop level. I could see Canari women in their colorful cholas, washing their clothes on the riverbanks, turning the rocks into a rainbow of colors. God, it felt good to be back in the cool, crisp mountain air of the city with the red tile rooftops and the beautiful old colonial buildings.

The Spanish founded the Ciudad De Cuenca in 1557. It was nicknamed, "Fifteenth century Spain under glass," a city where cobblestone streets amble by colonial-style houses, with their white adobe walls, metal balconies, and great double doors that offered privacy and safety to those that lived behind them. At 8300 feet above sea level the valley of Cuenca was chaperoned by towering mountains on all sides. It was one of the few South American cities with pure drinking water, clean air and a sense of peacefulness about it. "Muy tranquilo," the Cuencanos bragged of their home.

Cuenca was a sacred city. The great Canari civilization had flourished in the area around 500 A.D. Then came the Inca invasion. The temples and beauty of the previous civilization was destroyed and replaced by Inca palaces and their magnificent "Temple of the Sun." History once again repeated itself when the Spaniards arrived and the cycle of destruction and rebuilding continued with Catholic churches. Cuenca was a city whose foundation was a mixture of legend, indigenous culture, and Spanish domination. It was a city where time seemed to stand still. It wasn't difficult to imagine that darkened colonial archways were the fortress of mysteries from the past and the present. It was the Inca Gate of the Puma, Puma Pungu, and the gateway to the Oriente and the mountain village of Nabon.

The Hotel Cuenca was home while we waited for the local tinsmith to finish building the dredge's sluice box. The modern building, by Cuenca standards, was a tourist destination that was tolerable if you were there for only a couple days or if you sought refuge from the afternoon rain showers. Every afternoon between 2:00 and 4:00 the pitter pat of the cold rain splashed on the cobblestones. The rain always heralded the demise of electrical power, which ended our daily gin game. There was seldom any hot water even when the power was on so showering at any time was a sobering experience. Nights were spent flipping and flopping in discomfort in what Bob called, "Damn hammocks with humps in the middle." Comfort and privacy were absolutely non-existent. Our room was glass enclosed and faced an open stairwell, a location that amplified all the noise coming up from the bar and dining room.

Mornings were spent shopping at the Indian markets that flourished in and around the city. We bought bright yellow water containers, lethal looking machetes, pots and pans for cooking, rope, and jerry cans to haul the gasoline for the dredge. Our tents and dredge parts, including the motor had arrived from Guayaquil and I concluded negotiations for a car and driver to take us to Nabon. Our driver agreed to be available to come and get us in the event of any emergency. The Land rover was vintage 1960's, built like an armored car. The lease provision for the car was typical Ecuadorian; if a tire, transmission or engine blew, I had to pay.

Land Rovers and drivers were hard to find as many of the owners of these expensive vehicles refused to allow them to leave the city. The roads, if you could call them roads, were in terrible shape including the Pan American highway that linked Cuenca with Quito to the north and Peru to the south. The Pan Am was funded by the American taxpayers and was heralded as a critical link and final solution to Ecuador's progress and commerce. It was their way out of third world country status. The problem was that it was never finished. What was supposed to be a paved highway wound up with large sections of rutted, dirt road that were almost impossible to travel during the rainy season and the Pan Am to Nabon was no exception. Nevertheless, we were ready to go and the only thing holding us back was the tin man.

We calculated that it would take a small army of about fifty-five Canari to haul all of the equipment, including dredge, engine, sluice box, tents, food and one hundred and ten gallons of gasoline and fifty gallons of pure water for drinking and cooking. I learned from past experience that forty-pound loads were sheer torture on the backbreaking trail to Infiernilios. Therefore, we would lighten the loads carried by each man by adding more cargo carriers. Sounds simple, but it was far from that. We estimated that it would require twenty-five cargo carriers for just gasoline and water requirements. Each five new men hired for the journey required adding another carrier just to haul food for them and him.

Compounding these new logistic problems was the necessity to select, interview and hire fifty people of whom we had no knowledge. Past experience taught me that we would need a good translator, someone we could not only trust but who would be willing to protect our flanks. I put out the word with the desk clerk at the hotel before we left for Nabon to set up our pending expedition.

Francisco and Carlos had been hanging out all morning at the telegraph office waiting for my arrival with their poke and someone named Roberto. As we pulled up to the dilapidated shack housing the telegraph office they came running out with big smiles on their leathered faces. I introduced them to Bob and paid them with a bundle of sucres that would choke a horse. Sucres was the only monetary exchange accepted in remote villages. Checks were an insult to one's intelligence as paper had no meaning and plastic was unheard of out there. Before long word traveled along the dusty streets that "Rubio", the blond American, had returned. The news was like a magnet attracting men, women, children and dogs, all making their way towards us.

The local priest, a young Spaniard, introduced himself and offered his assistance. He thanked us for giving hope to the people of his parish. We passed out our gifts, which included candy for the children and scarves for the women, but we hadn't anticipated this large welcoming and soon ran short. The priest sensed our embarrassing predicament and came to our rescue. He told the people that they must leave because the Americans had much to do.

Thankfully, we retreated into the telegraph building to have some private time with Carlos and Francisco, who were beaming with pride. I asked them to find fifty strong and trustworthy Canari who would be willing to haul equipment and supplies to the river. They both assured us they would do their best and before they left we told them we would send a telegram to set the date of our expedition.

Arturo, the jefe of the telegraph office and the village's only link to the outside world, agreed to keep us updated on weather conditions, but he cautioned us that we must make the journey soon. The elders of the village had warned him that there were indications that the rainy season would begin early. It was a point I heeded after our previous experience with unpredictable weather. Amid handshakes and pats on the back, we left, assuring our crew that we'd return soon.

We had traveled about 20 clicks down the Pan Am trail when we were caught in a horrendous cloudburst so violent that it forced our driver to pull to the side of the road. We sat in the Land Rover knowing that there was nothing we could do except wait until the rain subsided. The vehicle began to vibrate, slowly at first then more ferociously. We watched in horror as a derumbe, a landslide closed the road in front of us. The mudslide had blocked the road to Cuenca and we were stranded until a road repair crew arrived to open the road. In Ecuador this could take days, if we were lucky. Rather than wait on a mud slick road that was becoming more dangerous by the hour, we decided to back track and try to find better and safer refuge.

At dusk we located a small house and knocked on the door. A short stocky Indian man opened the door and when he saw us dripping wet, he immediately invited us in. In broken Spanish I explained our problem and asked if we could spend the night. He was obliging and friendly and then asked if we were hungry. We nodded and he quickly disappeared into what we assumed was his kitchen. In the meantime, I waded back to the vehicle to tell our driver that we had a place to sleep for the night. Surprisingly, he refused to leave the Rover and cautioned us to be careful.

When I returned to the house, Bob was sitting on the floor cross-legged and munching on a boiled egg with a hot cup of coffee as a chaser while our host was busy arranging our sleeping area by placing pallets on the floor. "Where's Emilio?" Bob asked.

"He wants to sleep with his Rover." I replied.

Bob shook his head in disbelief. "He's afraid, isn't he?

I nodded. "But afraid of what? He's either afraid to leave the Rover because he's responsible for it or it's because he's Spanish and afraid of Indians because of four hundred years of bad blood between them."

"I opt for the latter," Bob said. Our conversation ended when the Indian brought me my meal for the evening and asked if we were the Americans, working with the people of Nabon.

"My God, news travels fast in this country." Bob blurted out, almost choking on his egg. "Does everyone in this area know what we are doing?"

Ignoring Bob, I replied to our host, "Yes, we are the ones. But, how did you know?"

"Oh, simple," he replied. "The people in my area have heard the stories about Rubio, the American that appeared one day to save the village of Nabon. None of the people can remember your names, so they only refer to you as, El Hombre Rubio." Bob sat on the floor shaking his head and chuckling. Then we both started laughing when I mussed my wet blonde hair as if to substantiate his story. "If I'm El Rubio then you must be El Blanco," I said to my gray-haired friend.

The Swede slapped the floor between loud laughter. "They think you're a freaking savior."

Our host joined our boisterous laughter and the three of us sat on the floor totally amused at the ridiculous concept. Our host's name was Pepe. He was a simple man who raised chickens and pigs and sold them in various market places to make a living. He lived alone and had no wife or family, just his beloved Gato. Before leaving us for the evening he said that he was honored to have us as houseguests, and that we could stay as long as we wanted. "Buenos noches," he said as he slipped behind the curtain separating the two rooms.

A little later he peeked from behind the curtain and held up an old portable radio that he proudly hung on a nail on the wall. Pepe tuned to the Voice of America for our listening pleasure and waved goodnight. The room was sparsely decorated with old wood furniture and I guessed that it served as his living room and bedroom. The house was cold, but not unbearably cold, and the blankets and the pallets made a comfortable bed.

That evening Bing Crosby serenaded us with, "There's no place like home for the Holidays," while a large photo of John Fitzgerald Kennedy smiled down on us, alongside the compassionate gaze from the Virgin Mary. It was a surreal setting. Here we were thousands of miles from home on the eve of the holidays, in a strange land we didn't understand and in a home of a man that we didn't know. Outside our Spanish driver was freezing to death in his car because he elected to protect himself from the man giving us shelter. And to make things even more bizarre, I had inadvertently become the salvation of a small village whose population was looking to us to deliver them to a better station in life. Talk about pressure. What began with adventure and a normal gold venture had somehow transformed into a mission of mercy.

I blew out the candles and soon began drifting towards sleep. Almost there, in that in between state of awareness and deep slumber, Bob's nemesis from the past came back to haunt him. He jarred me awake with a loud, "Son-of-a-bitch. There's a rat in here." He jumped to his feet and grabbed his flashlight to hunt down the critter. His wild commotion brought a wide-eyed Pepe running into the room. Quickly sizing up the problem Pepe retreated behind the curtained-off room for a few moments and then returned with a tiny kitten that he assured us, would track the rat. The tiny size of the gato even brought a chuckle to my frenzied friend. We settled back on the pallets and tried to sleep as a marathon footrace took place between rato and gato. Who won, we didn't have a clue. But there was one thing we did know for certain; the rainy season was upon us and we would have to move fast.

When we arrived back in Cuenca, much to my surprise was a message from the hotel desk clerk informing me that he had found a translator. 'Phone this number and ask for Adriano', read the note. Not wanting to waste more precious time, I called immediately and set up an appointment through Adriano's mother.

Adriano arrived at the designated time. Both Bob and I were a little surprised by the fact that he was prompt but also by his appearance. He was nattily dressed in a black suit, which was complete with vest and a red striped tie. He was tall and muscularly built with strong arms and shoulders that were indicative of a body builder. His rugged face displayed Italian and Spanish features; black Spanish eyes that were closely set above a classical Italian nose. He appeared to be in his early twenties. We introduced ourselves and I invited him to have a beer but he politely refused saying that he did not drink or smoke. I commented on his excellent physical condition and he replied that he kept in shape by pumping iron, had a black belt in Karate and a brown belt in Judo. Bob and I glanced at each other; both of us were thinking this was too good to be true. But, then it got better.

His name was Adriano Vintimilla Ordonez and his family had migrated to Ecuador, centuries ago, from Vintimilla, their family village, in Northern Italy. Adriano, as he preferred to be called, was a third year law student at the University of Cuenca and while attending school he worked as an assistant Judge for the Municipal Court of Justice. His passion in life was education and exploration, but he added that between his studies and work, there was little time for adventure. He hesitantly admitted, as if hiding a personality flaw, that his curiosity got the best of him when Ernesto from the hotel contacted him and explained that two explorers from the United States were looking for a translator.

During the two-hour interview, Adriano listened intently to the details surrounding the Infiernilios project and agreed that we must act quickly because the weather could change at any time. He agreed to work with us on one condition: he would have the responsibility for our safety and insisted on completing a background check on everyone that we hired, including the men that had worked with me in the past. He assured us the investigation would only take a few days, providing he had a list to work from. I asked him when he could go to Nabon and he said he would make arrangements for us to leave the following day, in his families Land Rover. He shook our hands and started to leave, and then he turned and said, "Oh, by the way, if you are successful and need a place to sell the gold, my family is majority stock holder in The Banco Azuay."

After he left Bob and I had a couple of beers. Our goal was to find a good translator; now we had one, along with a protector, assistant judge, banker, driver and what I hoped would be a good friend. We hurried downtown to send a telegram to Arturo in Nabon. The ball was rolling.

The following day we entered the town-square of Nabon and Adriano muttered, "Oh my God." He knew it was going to be a long morning. The villagers had lined up at sunrise near the church with a prayer on their lips and with hope in their hearts that they would be the lucky ones chosen. Old men, young men, those with families and those without, all waited patiently in a slow moving mist. There were hundreds of them. It was heart wrenching to see so many desperate people asking for an opportunity to work. It was difficult to turn down those who had brought their small children with them, the kids with big eyes and emaciated look on their unwashed faces. They reminded me of some of the young Apache kids I grew up with. My stomach knotted when dozens of little old women tugged at my shirtsleeve, begging me to select their husbands or their sons for our work team.

By mid-morning we had already hired the eight full timers and the forty-seven cargo carriers. Adriano was busy getting vital information to run background checks. Even after we had terminated our selection process, dozens still hung around begging for work. The young priest again intervened and ushered them away saying that if this was successful there would be other trips to the river. We unloaded the jerry cans full of high-octane gasoline and the three tents. It was time to leave and on our walk back to the Land Rover, Bob noticed the tears in my eyes. He gently patted me on the back. "We've done all we can for now, Lee," he said quietly.

The drive back to Cuenca was silent. All three of us were lost in deep thought. Adriano focused on the road and Bob in the back seat stared out of the window at the rugged Andean landscape. I wanted to talk, but I couldn't find words to ease or erase my sorrow. I thought about those poor kids who had nothing, no education, and no hope for the present or future. My sadness for the children began to leave but it was being replaced by anger. Why didn't the government, the church, somebody, do something? Here we were in the 1960s planning to place human beings on the moon while this country lived in few centuries behind the rest of the world. It made no damn sense at all.

In the distance I could see the three massive blue domes of The New Cathedral, which told us we were near Cuenca. After eighty years the New Cathedral was still under construction. It was overwhelming: 100 meters long, 50 meters wide and 40 meters in height. The three-story altar was gold plated and the tons of Italian marble used in the construction came from the world-renowned quarries of Carrara, Italy. The white marble created a sea of reflective images to those that came to pray and ask for God's forgiveness. The bowels of the massive church were honeycombed with tunnels and burial crypts reserved for the religious elite. The mere presence of this colossal house of worship was overpowering. A local Priest told me that, "It represents a beautiful testimony to man's concept of the power of God." All I could think of was what if the money had gone to educate the people instead of creating a place they could seek absolution.

The following day, Adriano called and delivered bad news. Three of our crew selections had criminal records. We had hired two murderers and one horse thief and we would have to replace them. The Republic of Ecuador's justice system was based on Napoleonic Law meaning that an individual is guilty until proven innocent. Adriano, our aspiring law student, was adamant that we terminate the criminals. I would have been willing to look the other way concerning the horse thief, especially in the lawlessness regions of the Oriente and mountain villages like Nabon. But, Adriano's pragmatic rationalization was once a thief always a thief. Not wanting to waste any more time because of the narrowing window of weather, we'd have to terminate and replace them on the morning of our departure. I had a bad feeling about the timing of our decision but it was something we would have to live with.

Cuenca was still asleep as we navigated the packed Land Rover through the narrow cobblestone streets and across the Tomebamba Bridge, towards the road leading to Nabon. As I gazed to the north of the city, towards the cities of Canar and Biblian, I thought about the carnage of the great battle that had been fought in the area.

The Incas wanted this region for their great northern outpost and invaded with an army of four hundred thousand fighters, who attacked the Canari army of two hundred thousand brave warriors. They fought in hand to hand combat, using only spears, knives, axes and rocks. At the end of the day, sixty thousand brave men lay dead on the battlefield. Historians said the valley ran red with blood and that it took weeks to bury the dead. After the battle, the Incas realized that the price they paid in manpower was too great. A truce was negotiated and the Canari were invited to share power in the region. Their dignity and pride had survived through their determination to protect their land. Then the Conquistadors arrived and took everything from the great Canari and Inca. They lost their pride, dignity, land and their language, and in this corner of their once vast empire it showed the most.

Somehow it seemed fitting that we arrived in the church square at sunrise on a Sunday morning. The congregation of humanity awaited us dressed in their finest. It appeared that the whole village had turned out for our return and the young priest was ecstatic as he bounded up to our Rover. He welcomed us and then left for the church, because he was about to conduct the largest mass in the history of his parish. After mass we were unloading our equipment and supplies when he climbed on top of the Land Rover to address the large crowd. Adriano translated the priest's passionate pleas to his flock.

He stressed teamwork, love and understanding, and most of all cooperation and respect for the two Americans who had brought hope to the people of Nabon. He concluded his sermon on the rover, with "Vaya con Dios," and as if on cue the packhorses were led into the square. The sight of the horses created mass confusion among the cargo carriers, as they surged forward towards our equipment and supplies, jockeying for position and weight loads that were being distributed on wooden frames, in back packs and knapsacks.

While Bob, Francisco and Carlos distributed equal weight loads to the eager mob, I noticed Adriano in a heated discussion with the three men that were being terminated. One spat on the ground in disgust, another shifted his feet in a nervous manner, while, the stocky one with the scar above the right eye, was in Adriano's face. Adriano, at five feet eleven inches, towered over the short, stocky Neanderthal type, but the Indian was not backing down. The veins in Adriano's neck began to expand in rhythm with the discoloring in his face; I could see that he was about to blow. I made my way through the crowd to lend support when the confrontation abruptly ended and the three men retreated to a group of men standing nearby. After some ugly stares aimed our direction, they left the village square.

Adriano made eye contact and motioned me to join him. He reached into the glove box of his Land Rover and pulled out a leather shoulder holster that housed a .38 snub nose revolver and twenty extra rounds of ammo, "Here take this," he said quietly. "You might need it!"

He was still angry when we made our way back to the packhorses to assist our crew with the weight loads. Calm had been restored and we were ready to leave, when Ochoa brought over a young man who Ochoa said spoke English.

Excitedly, I asked, "Hi, what is you name?"

He responded, "Hi, my name is Washington and I speak English."

"You're hired," I said.

It was a rare joyous moment for the small village whose history of pain and suffering was apparent. Everyone seemed filled with joy and appreciation for El Blanco and El Rubio who had agreed to share their wealth. Everyone that is, except the three that had been canned.

The morning sun was high when we finally hit the trail, all one hundred and fifty of us. Wives, mothers, fathers, and children accompanied the fifty-five we had hired and the additions were self-appointed cargo carriers, carrying what they could to help out. One young woman was carrying the foot valve, a girl had a coil of rope, and children had bags of food. A festive mood stirred the air. Everyone talked and laughed while passing bottles of white liquid back and forth. I looked over at Bob, grinning from ear to ear. "This reminds me of my creditors back home," he said howling.

I yelled at Washington and he came running up to me. I asked him to find Carlos.

He smiled and replied. "My name is Washington and I speak English."

"Yes, I am aware of your name. Now please, go and find Carlos." I repeated.

He smiled once again while repeating, "My name is Washington and I speak English."

"Shit," I mumbled to myself realizing that was the extent of his English.

By the time we reached the Rio Shingata our expedition looked like a Moslem migration to Mecca without any religious overtones. People were scattered throughout the high plains. Half of them weaving and bobbing in an uncontrollable manor that had nothing to do with the heavy cargo, but was the result of the white lightning that had been passed out freely on the trail. What began as unity was headed toward chaos.

The Father's flock was still flying high by the time we made our first camp for the evening. Although most of the self-appointed carriers had already turned back, there were still a few that were too drunk to move. Bob did a quick head count; we had sixty-seven souls with us but over staffing was the least of our problems. Fights broke out between drunken macho men who challenged each other over whatever riled them. Fistfights were acceptable, we could break those up, but when the machetes came out all we could do was step aside and let the crazy bastards carve each other up. Finally, exhaustion replaced hostility and everyone bedded down for the evening.

The next morning we awoke to a nightmare! Francisco and Carlos informed us that nine of our selected cargo carriers deserted during the night. Eight said they were sick and one was cut up during a machete fight. In one fell swoop we had lost almost one-seventh of our scheduled manpower. Even the extra people we had counted had slipped out early and staggered back to their homes. We would have to double up some of the men's loads. One item that was impossible to double up on was the gasoline, which was in five-gallon containers. It was too heavy for a man to carry two five-gallon containers over a treacherous trail for two days. I had no choice. Thirty-five gallons would be covered with a tarp and left on the trail. Once we reached the river I could send men back to pick up the stored gas cans and bring them to base camp.

At first the men balked at double loads until I offered to double their wages, then everyone wanted a double load. I wondered what happened to all of the cohesive cooperation that was being displayed so openly and freely in the village square? I also wondered why had the cargo carriers left the expedition to return home? Was the sickness physical or mental or both? Was greed already rearing its ugly head? I tried desperately to make some kind of sense out of the craziness as we balanced loads and negotiated prices. The wrangling went on for the good part of an hour before we settled the differences and moved out, marching single file for the better part of the day before descending into the Valley of Darkness, where we camped the second night.

Under lamplight Bob finalized the fuel calculations for the dredge while I took inventory of the sizable bundle of sucre notes, brought along as payment for those that would dump their cargo at Infiernilios and then return home. Adriano and Arturo had worked out a logistical plan to supply us every ten days, so we were not worried about fresh supplies and we knew that we could send men back to retrieve the gasoline left behind.

Our only concern was the unpredictable weather in the cordilleras and why five men deserted our camp and the remaining men refused to discuss their departures. Compounding these headaches was the acrid smell of sulfur brought into camp by the rustling wind. Yana Cocha was indeed a depressing place that seemed to drain positive energy and create a feeling of despair and loneliness. Even the Canari despised this small body of water and the ugly black soil that surrounded it.

The next morning Francisco yanked open our tent flap. Struggling to shake off drowsiness, we were informed of more bad news; seven more men had left before daybreak. I demanded to know what the hell was going on. Calmly and slowly, Francisco whispered his suspicions. One or more of the men were causing trouble, threatening the others if they did not abandon the expedition. The innocent men feared retribution and were afraid to talk. He promised to circulate among the men to try and draw out the truth. We packed up and again we left nonessential items stored on the trail and doubled up those still carrying single loads. If we could just get to the crest of the Cordillera Espadillas we would be in good shape and from that location we could run shuttles to bring up the remaining supplies.

As we struck camp dark storm clouds rolled in hanging low over the valley. Then tragedy struck. Bob went down; he was deathly ill, as was Ochoa, and several others. All were moaning and heaving up breakfast in uncontrollable spasms that made me think it might be food poisoning, but we had all eaten the same thing. Then I remembered that Bob and the others had asked to have their canteens filled with the pure water brought from Cuenca. I spread a dry tarp in the wet grass and tried to make Bob as comfortable as possible, while Carlos attended to the others.

Francisco pointed to the young Indian boy that had been assigned to carry a five-gallon container of fresh water as the suspect. I motioned for Washington to bring me my pack while keeping eye contact with the young boy who acted nervous. I pulled the .38 and its holster from the pack and clipped them to my belt. Silence fell over the area. I approached the frightened boy who was beginning to tremble. He dropped the pack frame that housed the contaminated water. Carlos sensed the danger, ran over and got between us, and then reached down, picked up the container and unscrewed the cap and poured some liquid in the cap. He thrust the liquid towards the boy, "Bebida", he shouted angrily.

The boy turned away refusing to drink and slumped to his knees pleading with anyone that would listen that he was innocent.

Pouring the tainted water in his hand, "Malo, muy malo," Carlos moaned and then threw the cap at the boy striking him in the chest. Shaking with fear, the young Indian boy could no longer continue the charade. He broke into tears and confessed that he had been ordered to dump the pure water and fill the canteen with the lagoon water.

"Who ordered you to do this?" Carlos demanded. Trembling, he pointed to a stocky Indian who was resting on his poncho.

The drizzling rain failed to cool my rage. I wrapped the palm of my right hand around the butt of the holstered pistol and walked aggressively towards the accused that quickly scampered to his feet and began apologizing in a rapid flow of high-pitched Spanish. Approaching the stocky Indian, I recognized him as being the one that gave Adriano a hard time in the square. He was one of the murderers that we had terminated. Somehow he had infiltrated into our crew during the confusion on the trail and had blended in as an innocent cargo carrier.

Like a cornered animal he began a frantic denial. He assured me that he was unaware the lagoon's water was bad, insisting he was sorry and made a mistake while nervously glancing at the weapon on my belt. His lack of emotional regret and the look in his eyes told me he was stalling which made me even angrier. In his face now, I told him how stupid he was. Carlos edged between us and took my arm and led me away, saying that the man was not worth going to jail over.

Carlos reprimanded the thug and told him he was on a short leash and not to stray out of sight. I then walked over to check on Bob, who was now lying on the tarp, propped on one elbow. His complexion that was a mixture of colors that varied from pale to greenish cast. "How are you feeling, amigo?" I asked quietly.

"Like shit," he replied weakly.

I assured my buddy we were headed back then told Carlos to stop everyone and start unpacking. The expedition was over!

The cargo carriers were spent from carrying double loads and those sick from the bad water couldn't move; it was an easy call to make. I made up my mind that I would walk out, get horses and return to the ridge above Yana Cocha. It would be an all night journey, but two men, traveling light, should be able to reach Nabon by early morning. In the meantime, Carlos would be in charge. He was instructed to see to Bob's every need and to move everyone up the pass at daybreak.

Carlos ordered the men to set up the tents for those that were ill and build lean-tos made out of grass for the rest. I placed my canteen of fresh water next to Bob and told him not to worry. "I'll fire off two rounds when I reach the crest of Yana Cocha." I patted him on the shoulder and said good-bye. He was too weak to respond. I rummaged through my pack discarding anything with weight that was not a necessity and yelled at Washington to do likewise. We had to leave. Darkness would be on us within the next two hours.

Evening shadows were crawling along the ridge when we began the grueling climb up Yana Cocha, sweating and puffing in our leather ponchos that kept the rain at bay. Midway up the mountain rain began to fall, light at first then heavier and soon a deluge of running water was cutting into the old trail. We climbed higher and higher, pausing every so often to regenerate strength and breath. I followed close to the lanky Canar occasionally being thumped by his foot or getting his heel in my chest when he slipped on the muddy path.

We came to a point of the trail with a shear drop. It suddenly narrowed, forcing us to place our shoulders and both hands against the wall of the cliff to steady our balance. Trying not to look down we edged our way around the slippery slope only to run into another wall of dirt and rock. Both of us were covered in mud from the falls we had already taken on the dangerous trail and before long we welcomed the refreshing barrage of rain that cleansed and revived.

Reaching the top of Yana Cocha, we collapsed on our backs between two large boulders, which afforded us a windbreak from the strong currents. The moon, which had periodically been our light bearer, had disappeared behind boiling black clouds. We started to shiver as icy rain began to fall. After giving my tired legs and depleted lungs a chance to regroup, I staggered to my feet, raised the 38 above my head and slowly squeezed off two rounds. The wind was so strong that I questioned whether the boys at the lagoon heard the gunfire. The rain was cold and hard, peppering our skin like a wet sandstorm.

Retreating behind boulders, Washington and I sat in silence eating beef jerky, trying desperately to convince our tired bodies to move on. The rain refused to let up and with the moon hidden by the clouds we would have to rely on a small flashlight to guide us across the ridge. I organized my thoughts and then used the boulder to steady myself as I strained to get upright. I took one step and fell flat on my face with my legs trembling uncontrollably. My body was rebelling and my mind was edging it on. Washington was now laughing hysterically as he bent over to help me up. Grabbing his hand, I pulled him off balance and he went tumbling into the mud with me. We lay there covered in mud, laughing so hard that it was difficult to discern tears from rain running down our faces. We looked like two kids after their first mud fight.

Then reality caught up with me. The real world was down below where Bob and my men were struggling to stay alive. We were only fighting the cold and fatigue. The rain turned into a full-fledged downpour stinging our faces and soaking our boots with muddy water now standing four inches deep in the trench we called a trail. Guided by the narrow beam of the flashlight, we slipped and slid down embankments and troughs, covered with the thick, sticky mud of the highlands. Our bodies were numb to all feeling and our minds numb to logic. When we did stop to rest, it was in a mud puddle. But that didn't really matter. The only thing important was a moments rest that my wet, tired body craved. I hoped that we would reach the Rio Shingata before it overflowed its banks. If we arrived too late we'd be stranded and unable to reach Nabon.

As we pushed our minds and bodies forward, the rain started to let up and that boosted our spirits momentarily until the flashlight grew dimmer and dimmer until the batteries died. With no moonlight we were forced to rely on our own senses to guide us. Taking off my thick leather belt, I placed it in Washington's hand. We used it as an umbilical cord to link and support each other on down hill slopes. When one would start to slip or fall, the other would plant his feet or hand to try to keep the other upright. Sometimes it was effective and sometimes it wasn't, but it acted as a powerful psychological tool and was our lifeline when we took tumbles together.

Bruised and battered we continued the decent into darkness until we reached the east bank of the Rio Culebrillas. The small tributary we waded yesterday when the water was only knee high, was now muddy brown water that was waist high. Grabbing the belt to connect and steady us we plowed in and struggled to the opposite bank. My body was so numb to feeling that another ice bath had no effect on my short-circuited senses. It was only when the cold wind grabbed at me that I winced.

Finally, we reached level ground and, in the blackness of night stumbled into a large boulder. The obstacle became a godsend that offered us a chance to lean on something instead of squatting in running water. We took our first smoke break and downed swigs of cool invigorating water as we gazed at a wondrous sight. The moon peeked from behind the clouds and became brighter and brighter, as the storm moved to the south. My legs were trembling so bad I was afraid they were going to cramp up so I spread my poncho on the muddy ground flopped down and with both hands began messaging them. Washington did likewise and we sat there trying to regain our strength while gazing into the moonlit sky. I asked Washington if he was glad he conned me into joining the expedition. There was a long pause, before he replied. "Si Senor Lee."

"You are a master bull shit artist." I replied in my mountain Spanish mocking him with, "My name is Lee and I speak English." We both needed a laugh. Humor was the only thing we had going for us.

Our energy returned and we continued on feeling strong until we reached the freezing waters of the Rio Shingata. The task before us was disheartening. We faced a raging current of a river that was rising at an alarming rate. We tightened our grips on my belt and plunged in. The freezing river water took my breath away but at least it was washing off all the gooey mud. We inched along, bracing one another and fighting against the downward pull of the angry river. Stepping slowly and cautiously, we finally reached the far bank and crawled up its muddy incline. We were soaking wet and freezing cold. My teeth chattered to the point my jaw hurt.

Peeling off my mountain boots and soaked socks, I suddenly realized that my brain and lower body were functioning again, the shock of the cold river had brought them back to life. I hoped that this was the stimulus required for going that extra mile. Washington and I had to keep moving to generate enough body heat to keep the wet clothes from freezing to our skin and to keep us from going into shock. I put on my only pair of dry socks, laced up my boots and grabbed his arm and yelled, "Vamos." He shook his head as if to say no. Washington was struggling, he wanted to stop, rest and sleep. "No es possible." I yelled above the roar of the river.

He refused my command and lowered his head shutting me out of his life. I was infuriated and let him have it. "You're a weak bastard. I thought Canari men were tough. You aren't tough if you're going to stay here and die." He refused to talk. "Okay then, I'll go without you."

With that final outburst, I took my belt and threw it at him, hitting him on the foot as he sat staring blankly at the ground. I thought my anger and insults would piss him off to the point that he would join me, if nothing else to prove that I was wrong about him. But, he just sat there with a distant blank stare, so I started off on my own.

The bright moonlight was guiding me past recognizable landmarks. I knew that the village of Nabon was only four to five hours away. I also knew that the only way I was going to make it back was to keep my body heat churning, so I ran, stopped to do jumping jacks and walked at a brisk pace. Soon my body temperature was normal, but my thoughts were becoming foggy and my legs were like dead weights. "I can't go on," I heard myself say as I literally collapsed into a clearing of wet grass. I lay there unable to move; my eyes slowly began to close; my mind was beginning to hallucinate. I heard my grandmother whispering, "Son, son. Get up. It's time to come home now." I felt Perro standing next to me, pawing at my boots. I remembered Washington with his head down, lost in a blank stare of defeat and how stupid I had been for leaving him alone to his fate. I thought of Bob, ashen and possibly dying. Then I became weightless and blacked out.

I was drifting in a state of semi-consciousness when I heard my name being called and, as I struggled to regain full consciousness, I heard the voice becoming louder and more distinct. "Senor Lee, donde esta?" the voice asked.

Semi awake, I recognized Washington's voice calling for me. My first thought was, "Oh my God, I'm dead, and we were banding together to make our final voyage into the hereafter." But the pain in my legs told me that I was still in the physical and that I must be hallucinating again. Then I heard the sound of heavy breathing and weak footsteps and his voice asking where I was. Opening my eyes, I saw a shadowed form of a tall man coming down the trail towards me. I knew that it was Washington and that we were both still alive. My sudden emergence from the clump of grass startled him and he went for his machete then realized that he had finally found me.

We grabbed and hugged each other and vowed that we would stay together until we reached the safe haven of the village. Then we sat down and finished the last of the water and beef jerky, and tried to get our bearings. I flicked my cigarette lighter to check my watch. There were bubbles of moisture under the faceplate and the hands had frozen at 12:02, which was probably the time when we waded across the Shingata. As I sat there rubbing my aching legs I heard that voice again. "Vamos." It was Washington and he was telling me that he wasn't a weak Canari and it was time for me to get off my ass and move on.

For the next two hours, we stumbled along the trail talking up a blue streak, like two friends that hadn't seen each other in years. I discovered that Washington was twenty years old, had very little schooling and had been on his own since he was sixteen. He was the prodigy of a broken home, where his father, like so many of the poor Canari of his era, was an alcoholic and his mother an illiterate basket and rug weaver. He vowed to better his life when he heard about my expedition to Infiernilios. He knew that we needed translators so he borrowed an English book from a friend and went about learning English. Without formal training and guidance, it was too difficult and he gave up after learning a few catch phrases. He had bluffed his way on board and was still damn glad he did, even though he was suffering as much as I was. I admired his spunk and will to succeed. He sort of reminded me of myself, a hundred years ago, it seemed.

After hours of constant chatter from my Canari friend who was named after the American president, our conversation wound down. Mental fatigue had set in once again. Our minds were demanding that we shut down, take a breather, maybe even a short nap and then we could regroup for the final push. A conspiracy was underway between the mind and body, with the mind tying to trick the body into thinking we could lie down for a few minutes and then continue. I knew that if we succumbed to this the body, through physical exhaustion and pain, would eliminate the mind over matter formula that was pushing us onward.

This deadly mix was more difficult to avoid by the hour and the only solution to keep it at bay was constant conversation. So I started a constant dialogue with my hiking partner, talking about anything that came to mind and throwing in questions to keep him alert and responsive. I remembered that the Mexican nationals that were hired by the Southern Pacific Railroad to lay rail at Peridot wanted me to teach them how to count in English, so they would not be cheated when they cashed their railroad checks. I asked Washington if he could count to one hundred, in English. He said he couldn't. So I started an English lesson stumbling through the darkness on what seemed like an endless trail.

"Uno, one, dos, two, tres, three, quatro, four, cinco, five, seis, six, siete, seven," I continued all the way to one-hundred, speaking the numbers slowly, in cadence with my stride. Over the next couple miles he was starting to get the hang of it. His exuberance was slowly returning.

As we climbed a small knoll bathed in moonlight my thoughts returned to Bob and the crew. Were they okay? Was Bob still alive? Would we ever get to Nabon? What if we were too late? I paused for a moment giving my body a breather. We didn't dare sit down, so we leaned against one another, back to back, breathing in the crisp cool air. We were quietly lost in our own thoughts when we heard the sound of dogs barking in the far distance. "Nabon!" Washington cried out. Thank God, for dogs, I thought, they were truly man's best friends.

We stumbled through the deserted village with a dozen mongrels snapping at our heels and made our way to the telegraph office where Arturo slept on a cot in the back room. Banging on the door we rousted him from sleep. He opened the door and stepped backwards, shocked by our appearance. "Madre de Dios," he cried.

We were so tired we just stood leaning in the doorway, numb to movement and conversation until he pulled us in by tugging on our wet clothes. At first I was afraid to sit down because my mind had been conditioned to stand for the sake of survival. It took a while to realize that we had made it and when I accepted that fact I slumped on the plank floor with my back against the wall.

We drank glass after glass of cool water and Arturo stared wide-eyed at us shaking his head in disbelief. We filled him in on our aborted expedition as I tried to stop my legs from shaking by stretching and rubbing them. They had cramped up so badly it was impossible to control them. Washington and Arturo left to find the owners of the horses, saying they would return within the hour. I glanced at the clock on the wall; it read 6:14 A.M. We had been on the trail for fourteen hours. Then I passed out.

I was having another one of those weird dreams or maybe a hallucination, whatever it was, it seemed to be happening in real time! There was a large talon above my face, inching closer and closer and there wasn't anything I could do about it. I was on my back looking up, frozen by fatigue and fear. I heard a woman's voice saying in Spanish, "Quire comer?" Was the voice asking the prehistoric bird if it wanted food or was the voice asking me if I wanted food? The questioner demanded a quick answer and my half-closed eyes popped wide open. The voice belonged to a large woman that I recognized as Francisco's wife and she was holding a live chicken inches from my mouth. "Yes, si," I responded weakly. "I would love some food."

She nodded in response. Handing the chicken to her daughter Francisco's wife aggressively approached me with a determined look on her face. Without any hesitance she began unbuckling my belt and undoing my fly. I thought I was really dreaming except that I could feel her fingers on my skin. As I rose in self-defense, she pushed me back down and said, "Limpiar." Again I tried to steady myself but she pushed me down again saying, "Si, lavar." Then it donned on me what she wanted. My clothes were filthy, and she was going to wash them, come hell or high water.

The sun was in the west when the rescue team arrived carrying Bob and the others who were too weak to move under their own power. The healthy members of the team had hauled back what they could but I didn't even want to guess what was left behind on the trail. I rushed to Bob and helped him dismount. Although he was still in bad shape he gave me a partial smile and staggered into the telegraph office. He slumped against the wall, a man that was moments away from total collapse.

Under the dim, bare bulb hanging in Arturo's office, I paid our men the money they were promised and gave those who had stuck with us a bonus and heart felt thanks. They were thankful, yet sad about the way things had turned out. It was disheartening to watch them file out one by one with heads down and their hopes and dreams shattered by a deceitful few.

Washington was one of the last to leave, but before he left, I grabbed him and gave him a bear hug wishing him the best and telling him not to ever give up on himself or his future. He thanked me and disappeared into the night. I never saw him again. He and many others I had developed a camaraderie with moved on in a constant search for employment at a dollar a day, and like my Grandfather said, "Be damn glad to get it."

Carlos and Francisco were still sitting on the floor and in no hurry to return to their previous life. Both were depressed. Their dreams evaporated in sulfur at Yana Cocha. I wanted to cheer them up, so I told them that if we came back for another expedition they would be guaranteed a job, because without them it just wouldn't be the same. My words seemed to give them a sense of relief. They both stood up and shook my hand and I told them I would pay extra if they retrieved the equipment left on the trail. They thanked me for giving them and the village a chance, saying that was all they ever wanted.

Standing in the doorway staring into the blackness. I shuttered from the chill of night, wondering what the future held for them and their families when Arturo interrupted my thoughts. He informed me that my telegram had been sent and Senor Vintimilla had responded and would be arriving within the hour.

Everyone was feeling let down, including Arturo. He would have played an intricate part in solving the logistics problem between Cuenca and the Rio Infiernilios. He'd been excited about supplementing his meager income with the telegraph company and eventually buying a home for him and his family. He, like the others, was discouraged by the turn of events. I walked over to him and patted him on the shoulder and told him to have someone help him store the equipment in the back room and that I would send a truck to pick it up. I handed him three hundred Sucres for his expenses. I also asked him to keep us informed of the activity and whereabouts of the three criminals. He agreed.

Lightening flashes and the low rumble of distant thunder intermingled with my guilt for the failed expedition when Bob interrupted my thoughts. "Lee, I don't think I'll be doing this again, so you promise me something. Don't you ever go balls to the wall again? Go slow. Don't get pressured and most of all don't get caught up in all the savior bullshit."

I turned and grinned at my friend. There was nothing I could say. Then we heard the car pull in. It was Adriano arriving on Christmas Eve to pick up two battered and beaten friends.

Within the week we boarded Aerolineas Peruanas, in Guayaquil, for our return trip to the United States. In three months I'd lost thirty-five pounds and suffered from malnutrition, pain, dehydration and betrayal. My friend from Yuma suffered as much or more and sadly this was his last expedition. Bob Olson never returned to Ecuador. I was emotionally drained when I left this beautiful wild country and knowing that I had been the first "white man" to conquer the mystique of Little Hell was of little consequence.

Rio Infiernilios continued to remain elusive to the villagers of Nabon, reinforcing the legend of the mountain spirits and their disdain for those that came in conquest rather than reverence to collect those tears of the sun. One thing was certain. The tears left in Infiernilios from the Inca Sun God were from sorrow and not joy.

CHAPTER 3

MEJIA'S EMERALDS

Emerald laden matrix found in Rio Tutanangosa by Padre Crespi

"This will probably be the most important letter you have ever read. Since you left my country, I have been searching for new projects that you might have interest in. And last week, a man who lives in one of my Father's houses came to see me. He told me that on one of his recent trips into the Oriente, he ran into a woman with two children that were near starvation, feeling sorry for them, he gave them some money so that they could return to their home in Colombia. Because of his generosity, the woman gave him a book that had been passed down through two generations of her family.

"Now, Lee, this is where the story becomes interesting. This woman had been searching in the jungle for a fabulous emerald treasure and mine that belongs to her family. Her only clues to the location to this wealth are contained in her family's old documents, which include the last will and testament, letters and a map that is 87 years old. For

preservation sake, they have been bound together into this old book, which I have examined very carefully. This I will discuss with you when you arrive in Cuenca.

"Lee, I believe that this incredible story is true and that the documents I have examined are real. The reason for my optimism is this. Since coming into contact with the book, I have been investigating this affair and have uncovered evidence that does support the story, which I will share with you in person.

"In closing, I will only say that the man who owns the book is willing to share and is ready to go on the expedition. My friend, if God is on our side, and we can locate Mejia's emeralds, this discovery could completely change the political and economic structure of my Country.

"Tu Amigo, Adriano Vintimilla"

In the aftermath of my recent tumultuous expedition I needed a soul stirring reason to return to Cuenca, and the letter from Adriano was it. Adriano and I had stayed in communication with each other. He documented and informed me of the conspiracy in Nabon that sabotaged our second expedition to Little Hell. And although I remained disillusioned by the betrayal, he remained steadfast for future exploration and even retrieved, cleaned and stored my dredging equipment while he researched new projects and kept me informed of the political issues in his country. He was concerned about the assassinations of Martin Luther king and Robert F. Kennedy and the race riots erupting throughout America. We shared information on the political climates of our countries and the various aspects that entwined our lives.

Adriano was conducting investigations for the Department of Justice in Cuenca and I was employed as a private investigator for the Capitol Detective Agency in Phoenix, Arizona. Even though we had only known each other briefly we continued to establish a good relationship through our correspondence. He wanted to know how the Swede was doing and I kept him informed of Bob's improving health. His letters always tempted my sense of adventure but none could match, or even come close to the impact that this letter delivered. It rekindled the smoldering fire of adventure in my soul and lured me back to the land of the Inca.

The next three weeks were a total blur as I worked long hours on endless stakeouts, compiling reports and undercover work on specific cases while building up my war chest. My lunch breaks and spare time was spent in the Phoenix library boning up on the history and background of precious stones. I planned to be armed to the teeth when I did return to Ecuador.

I learned that the name emerald is derived from the Greek word smaragdos that translates to "green stone". Nature is the mid-wife in the birth of the beryl family of which the emerald is considered to be the most precious. Deposits were found in or near pegmatite veins and mining was almost exclusively extraction from host rock. Significant deposits were to the north in Colombia, especially the Muzo mine northwest of Bogota. That mine produced the finest quality emeralds in the world. The green of a quality emerald was incomparable to any other stone. Unlike diamonds and rubies, emeralds were not abundant. The technical information was interesting, but the legacy they left on mankind was profound.

The history of the deep green stones was one of intrigue, violence and greed. At the time of the Spanish conquest of South America, the conquistadors stole thousands of the precious stones from the Incas. Emeralds were coveted over other precious stones. The "green fire stones", as the Spaniards called them, were used to adorn the golden idols in the temples and palaces of the Inca aristocracy. One of the largest was found in the forehead of their goddess of creation, Illa-tica, at the Temple of the Sun in Quito, Ecuador. It was said to be as large as an ostrich egg and was a magnificent deep green color. Historians said that this stone came from an emerald mine located in the heart of the Ecuadorian jungle but the Spaniards never found the mine.

The Bibliotheca National, in Madrid, Spain contains volumes on Pizarro's conquest of the Incas. The monk, Sanchez, one of the friars who accompanied Pizarro's expedition, wrote of seven large emeralds that had been given by Atahualpa, the great Inca King, to the conquistador shortly after he landed his ship in a small port in northern Peru.

The monk wrote that Pizarro had turned to him after the delegation left and said, "These I will send to our King as a gift of my undying esteem, but to you, padre, falls the task of learning the source of these magnificent stones. I want not seven, but seven hundred, yes, seven thousand, because in these," he said as he tapped the stones with his forefinger, "is the real treasure of the Incas. Give me their emeralds, padre, and you can have their gold!"

Although Sanchez tried desperately to learn the secret of the hidden Inca emerald mines, he failed miserably. As months passed and the savages rejected religion, the Spaniards turned to hideous forms of torture as a way to extract information. They were quartered, boiled in oil and lanced. Many died, and those that survived still refused to divulge the secret locations of the mines. All that the conquistadors could learn was that the mine lay deep within the impenetrable jungles of Ecuador.

I poured over historic records of emeralds found in South America and began to understand the significance and urgency of Adriano's letter. This could be a monumental discovery. Of course, everything depended on deciphering the contents found in the last will and testament of a man named Mejia. The appeal was immense. One bag of green fire could be the equivalent in value to a ton of gold. The logistics of transferring emeralds versus gold in a lawless land was ideal for the security conscious. I was relieved to know that the Mejia project did not require purchasing dredges, diving suits, equipment and all the other supplies that go hand in hand with alluvial gold mining. I could easily keep this expedition under the radar and not attract unwarranted attention providing security and confidentiality could be maintained while we developed the project. The effort seemed worth the gamble.

I arrived on September 6, 1968 in the old colonial city, Cuenca, whose motto was, "first God, then the Virgin". Adriano met me at the airport in his family's Mercedes and we began our slow crawl over cobblestone streets into the city teeming with market day crowds. Canari men strolled through the square adorned in red or black ponchos that framed their characteristically braided hair. They mingled with the women in long woven skirts; colorful wool shawls, and Panama hats adorned with long ribbons; all blending together with the modern day dress of the Cuencanos.

We were near the Hotel Cuenca when Adriano informed me that he and his family would be honored to have me stay at their home. They had a spare bedroom and besides the food and hospitality would be much better than the hotel. He must have remembered how Bob and I always complained about the living conditions there. It was an honor to be invited into one's home in Ecuador and I accepted wholeheartedly. The walk to the residence on cobblestone streets was like returning to the 16th century, with the exception of swirling exhaust fumes and honking horns from the industrial revolution that seemed to conflict with Cuenca's historical past.

Cuenca buildings had two sets of numbers. The first, before the hyphen was the block number; the second double number was the building number. Thus, the residence at Calle Borrero, 7-35, was the thirty-fifth house on block seven of the street. Like all the other old residences in the center of the city, it shared a common wall with the neighbors to the side and rear. The two-story home was over a hundred years old and was bounded on the east by the Central Bank of Ecuador (Banco Central). Across the street in front, was the ancient colonial La Concepcion church founded in the 16th century.

The entrance to the Vintimilla residence reminded me of a sturdy fortress with massive oak doors that were always open during the day and locked tight at night. Ten feet behind the doors was the second line of defense, an iron fence and gate that was electrified and equipped with an alarm system. It had proven to be formidable during revolutions and times of unrest.

The ground floor contained his father's office and library, the maid's quarters and two large storage rooms that were filled with firewood and miscellaneous household items. A double staircase led to four bedrooms, the kitchen, dining room, enclosed patio, another study and the living room. There was no television room because the electronic revolution had not yet arrived in the city. A staircase led from the patio to the roof and Adriano's exercise and weight rooms. The kitchen contained two stoves that sat side by side: one electric and one wood burning. I was told that when there was a revolution, which was often, the electrical power was always cut first and when that occurred they would switch to the wood-burning stove. The fact that power always went off in the afternoons when it rained also factored into the equation. Apparently the dam outside the city was low on water, which I interpreted as a good omen.

After the quick tour, I was shown my bedroom, which in fact, was Adriano's. He had moved his belongings into the guest bedroom feeling I would be more comfortable in his digs, which was near the patio and dining room. After unpacking I took in my new surroundings. The room was narrow but long and contained various mementos from Adriano's youth. There was a small desk in one corner and a large closet in the other. The bed looked firm on its heavy hardwood frame, a far cry from the beds in the hotel that the Swede had referred to as, "hammocks with humps in the middle".

I joined Adriano and Alfredo Berezueta in the upstairs study and tried to get as comfortable as possible in the straight back, hard wood chair. Senor Berezueta was dressed in a rumpled dark suit, with a dark tie that was haphazardly knotted. He appeared to be in his late fifties, was of medium build and sported a small mustache on his tired face. We exchanged smiles and nods. I fidgeted with my reel-to-reel tape recorder while he studied an old book lying on the desk, as we waited for Adriano to finish a telephone conversation. I gazed around the old, but ornate, high ceiling study. On one wall was what appeared to be the families crest and on the other wall, the family shield. Two large portraits of a man and woman dressed in early 19th century clothing, stared down at us. I assumed they were relatives. I was examining the family crest when Adriano finished his conversation and formally introduced me to Senor Alfredo Berezueta the man who found the last will and testament of Raphael Mejia.

Alfredo was a poor prospector who lived in Cuenca in one of the many homes owned by the Vintimillas. His wife worked in the local telegraph office and supported his dreams of finding El Dorado. When they had a few extra sucres he would head for the Oriente to search for gold placer. Adriano's father was sympathetic to the family and allowed them to live rent-free in one of his houses. Senor Berezueta owed the Vintimilla's his loyalty and that was why the Mejia project wound up on Adriano's doorstep.

For the next hour I directed questions to Berezueta, beginning with how he had come in contact with the last will and testament. He said that he was in the Oriente searching for gold when he came in contact with a woman and two children. They were in very bad shape, suffering from exhaustion and malnutrition. They hadn't eaten in days. He gave them some money for food and because of his generosity they told him they were searching for their family's treasure. Hardships of the search had taken a toll. The woman couldn't pursue the quest any further and offered him the book. He paid three hundred sucres for the book along with a promise that he would give them a share of anything found. Early on he had attempted some exploratory work in the area on his own but had run into great difficulties. He lacked the necessary money for cargo carriers and was overwhelmed by an uncharted jungle. He turned to his benefactor, the Vintimilla family for help.

In order to get to the emeralds Alfredo insisted that we would have to go through the Jivaro settlement of Sucua and since the Jivaro were suspicious of all outsiders, whom they called colonials, we would have to set up a good front to cover our true intentions. He suggested a hunting expedition for jaguar that was abundant in the area. He asked if I could make the necessary arrangements for the heavy firepower that would be needed. I assured him that I could although I knew that it would be difficult to bring guns into a country that strictly forbade any weapons beyond the twelve gauges and twenty-two-caliber variety. Then again, if our lives depended upon smuggling weapons for our safety and well being, I would find a way.

While Alfredo spoke with conviction on the merits of the project, I kept glancing at the old hardbound book and thinking about the dashed hopes and dreams of all of Mejia's family members that had searched for, but failed to locate the emeralds. Eighty years was a long time to maintain an elusive dream, but maybe my judgment was not being very fair to all those lives that had been affected. I had learned that everyone's personal search for El Dorado is much different than the next persons and time is of little consequence unless it has to do weather.

One of the Vintimilla maids served us in the dining room. I was lost in thought and picked at my arroz con pollo. My thoughts were on why Mejia never went back to what appeared to be the greatest emerald find since the Spanish discovered the Muzo outcropping? Mejia found the emeralds and then he lost them. How was that possible?

Raphael Bollanos Mejia was a Colombian who like hundreds of others at that time, had labored long and hard for the Quienera Company in Quito, Ecuador. The Quienera Company ran a harvesting operation in the jungle of eastern Ecuador. The workers were called cascarillas, which was a name derived from the substance that bore the miracle drug for malaria and yellow fever. In their constant search for quinina, (quinine}, a bitter, crystalline alkaloid, extracted from cinchona bark, they pushed deep into the uncharted territory of the Ecuadorian rainforest. Mejia began as a locator for the Quienera Company and graduated to cutting and storing the bark. Eventually he became one of the camp's hunters that helped provide additional food for the workers.

It was in this setting in 1881, that Mejia's life changed forever. Mejia had been tracking a large black bear all afternoon, hoping to get close enough for his old vintage shotgun to fatally wound the animal. Finally cornering his fierce adversary on a ridge above a landslide, he fired both barrels killing the bear instantly. When the bear fell, it started tumbling and tearing out chunks of vegetation and loose rocks, before it finally came to rest near a small river. While dressing out the dead carcass Mejia noticed a green reflection sparkling in white quartz in the afternoon sun. On closer examination, he noted, "white stones with green in them."

According to the old frayed documents he was unaware that he had found the mystical green fire. He thought it was green quartz or tourmaline, both of which were extremely deceptive to the poor soul searching for emeralds. Mejia knew that if it were tourmaline he would have more money in this one discovery than from a year of backbreaking work. If his discovery were emeralds, he would be rich beyond his wildest dreams.

Mejia was curious and meticulous. He made mental notes of the terrain and then used his machete to chop and hack at the quartz rock until the blade became dull. Then he used the stock of his shotgun until it broke. He had separated as much green from the white matrix as possible and estimated that he had tres arrobas, or approximately seventy-five pounds of mixed gemstone and host rock. The sun was hanging low on the horizon and he knew he must hurry to reach the safety of the Tambo Santa Rosa.

The Tambo was a large house with a rock foundation that was built from cedar trees with a roof of palm fronds. This was the cascarilla's living quarters in the dense jungle. It was also a safe haven at night when the big cats came out to search for food. The tambo, or tambu, which takes its meaning from a Quechua word, was a safe house located near the confluence of the rivers Santa Rosa and the Tutanangosa. Thus, it carried the name for the house that sat near a river named the Santa Rosa.

Before reaching the tambo, Mejia's heavy load of seventy-five pounds of stone became too much for him to carry. He decided to hide the tres arrobas and only bring into camp a few samples of the green stone. He located an odd shaped rock that he described as, "a big stone, open in the middle," where he stashed all but a few specimens.

Entering the way station he took the fresh bear meat to the cook and then searched for the camp foreman, Senor Barrana. After telling his boss the story of how he found the stones, he gave him one sample, a one-inch piece of deep green, and asked him if he would arrange for an analysis of the unusual material. Barrana promised Mejia he would send the stone to the Quinera office in Quito and report back to him when he had some information. Mejia in the meantime went back to work, knowing full well that it might take months before he received any details.

Months passed before the analysis report with the emerald sample arrived by sailing vessel from the Sempres Mollina Laboratory in Paris, France. Word was sent to Mejia that the specimen was high quality emerald! This exciting information prompted Mejia to turn over another sample to Barrana to be sold in Quito. His trusted friend went directly to Dr. Miguel Arroyo, the Director of the Quinina Company, who had made the initial arrangements for the analysis.

Dr. Arroyo was excited to learn the two specimens were for sale and called in Henry Perring, a British agent who represented the Quinera Company. Perring read the analysis report and purchased the emeralds for three hundred gold pesos. This amounted to about six thousand dollars, a king's ransom in those days. Barrana accepted the payment and promised to deliver the funds to his friend.

Mejia was in a festive and giving mood knowing that before long he would return to his family in Colombia a wealthy man. While he patiently waited for Barrana to arrive in camp with his new found fortune Mejia accepted an invitation from an Indian friend. The following was translated from page six of his last will and testament.

"In those years, I was invited by Singusha, who was a chief of the Jivaro, to come and drink chica at his house. I told Singusha he should not

leave this great land, because here you can find a lot of cascarilla, rubber, gold and emeralds of great value. Singusha asked me to show him the emeralds of a great value. And because of his interest I took him to the Tambo Santa Rosa, which is a place where the quinina workers would eat, rest and sleep. Then I went to my bed and pulled out a sack, from under my pillow and removed some stones and showed them to Singusha. After I showed these stones to my friend, his interest continued to grow and he wanted to know more. So, I promised to show him the place where I found the stones.

"I know these savages very well and I know that they do not believe everything you tell them. They ask the same questions, over and over until you tell them the truth. So, I took him to the place I found the stones. First, I took him to a place where there was once existed a ranch, but it had been destroyed, these ruins are on the left side of the Santa Rosa river and close to the junction of the Tutanangosa. This ranch was close to these two rivers, just a few blocks upriver.

"These stones were in the Santa Rosa River, to the left following the current upstream that was near the Tambo Santa Rosa. Singusha had no interest in the green stones other than curiosity, as most of the Jivaro's that lived in that area was only interested in hunting and fishing. Singusha warned me that if these green stones did indeed have great value, to be careful. In the past, he witnessed what the yellow material had done to the mind's of those that came in search of it."

Unfortunately the Jivaro's warning had come too late. Before Mejia could enjoy his newfound wealth civil war broke out in Ecuador and Mejia was conscripted into the army and sent to fight in the battle of Guayaquil. He was fighting against the first president and dictator of Ecuador, General Ignacio Vintimilla. The year was 1881.

Adriano smiled as he got up from behind the old oak desk and walked over to an antique book shelve and pulled a bulky volume from its resting place. He fumbled through the pages that provided a glimpse into his family's background. "General Ignacio Vintimilla was a relative of mine and was president of my country between 1876 and 1883. The battle of Guayaquil was fought in

1881 the same year Mejia found the emeralds. Mejia's last will and testament is dated 1887. Mejia is telling the truth! I have not only checked out these facts that I have read to you, but also checked on the Quinera Company that employed Mejia and the Court of Records in Quito document that Mejia's employer was in business at that time."

As the evening wore on the fascinating adventures of Raphael Mejia came to life from my friend's translation of the brittle pages of Mejia's old letters.

"Civil war broke out in Ecuador, against General Vintimilla and in one battle, I was hurt. This battle was against General Salassi and I lost my leg. I also got hurt in the chest when we fought in the battle of Guayaquil..."

Now I had the answer to the question that was baffling me earlier. Mejia didn't lose the emeralds. He lost his ability to return to them. Based on what Alfredo told us about the jungle around Sucua, it was definitely no place for a woman with two children, much less a handicapped man with one leg.

"I tried to get back to the mine, but God was not with me. It came to pass that Perring also suffered from the war, losing a foot during an artillery duel between General Salassi and General Vintimilla. I would find out later that the large emerald and the other went to Henry Perring for three hundred gold Pesos'. Perring sold my emeralds to a British Foreign Minister, named Hamilton and another British man named Wilson. After this transaction, Wilson and Perring entered into a partnership to do what ever was necessary to steal the mine from me. Perring came to see me and threatened to harm me unless I gave him the map. I had to raise my gun to him. He attacked me trying to get the map, but one of my friends Ricardi and some others came to my aid and that's why Perring was unable to locate the mine. I had to defend myself with two-hundred workers."

The location of this event was not in Ecuador but at Mejia's home in Colombia. After the war was over Perring used force to try to obtain the valuable map. The irony of this story was two handicapped men, one missing a foot and the other a leg, were fighting over an emerald mine that neither was physically capable of getting to.

After the revolution Mejia made a tragic mistake. He befriended General Salassi who was the losing general during the war and was under house arrest in Guayaquil. All of General Salassi's followers, especially those from other countries were exiled and Mejia, one of Salassi's closest friends, was sent back to Colombia.

His frustration was obvious in one of his letters to a friend in Ecuador, "*You must talk to the government and give me all the guarantees that I need in order to get me back into my country.*"

In the tone of his letter he does not consider himself a Colombian when he refers to Ecuador as "my country." He undoubtedly felt that his participation in fighting for the Republic of Ecuador gained him residence status through patriotism, but he underestimated the political intrigues that accompany a revolution, especially if one is on the losing side.

Mejia's attempts to return to the mine were an exercise in futility and out of desperation he tried one more time to enlist a sponsor for an expedition. On December 10, 1908, his frustration surfaced once again.

"*My dear friend, it seems that you are the only one who still wants to go into this affair. I don't know if the other people have retired because of political motives or other reasons. ...Rather than send you the map, I will go with you. But if you are not able to establish the association (agreement between us), then I will not go! ...The region is awfully rich, and if you have any doubts in forming a partnership, the facts will prove me right. The existence of the emerald mine is a fact that I will insure the success with my throat!*" (A Spanish expression that means if you don't have any success you can cut my throat.)

While Mejia was pledging his throat, a few months later his old friend, Barrana was trying to cut it. He betrayed Mejia. Barrana had received an interesting letter from a British mining and exploration firm that was interested in Mejia's emeralds. He replied to the letter in April of 1909.

"...I don't know the right place he found it, but I am sure it is in a ravine, close to the Tambo Santa Rosa. According to what Soto told me, who is a friend of El Sapo. I have been trying to engage this man, but I don't think he will agree."

The reason the British mining company contacted Barrana was obvious. Two years earlier Senor Barrana, in a letter to a friend in Quito, makes a very interesting confession:*"I used to work in a station of Quinina in the mountains and forests of Tubal. One of my workers was, Raphael Bollanos who had the nickname El. Sapo and a friend of General Salassi. He found two emeralds and one of the pieces he gave to me was about one inch large and the other one smaller. The large piece I sold to an Englishman named Henry Perring, who was an agent for the Quinina Company. My piece was sent to Paris, to Sempres and Mollina for analysis. Dr. Miguel Arroyo, who was the director and President of the Quinina Company, sent it. I was paid three-hundred gold pesos for that piece of emerald."* (Mejia was unaware of this letter up to his death and how the family obtained it from Barrana's personal correspondence is unknown.)

The emerald's twisting trail of intrigue and greed had finally brought us to the real culprit in this sordid affair. It was Mejia's boss and trusted friend at Tambo Santa Rosa, the same man that Mejia had entrusted with emerald samples to obtain an analysis on the stones. The story had become international. Through a friend at the Quinera Company, Mejia was able to follow the trail of the original two emeralds that had been given to his once trusted friend, Barrana. He knew that the first piece went to a laboratory in France, the result of which was an exciting analysis report that confirmed his find. This was the stone that Barrana sold for a king's ransom.

The second stone, which was smaller in size, was also bought by Perring and sold to a man named Hamilton who then sold it to Wilson. It was unclear whether Barrana received payment for that emerald or entrusted it with the Englishman in return for his participation in a side deal. Perhaps Barrana was in on the conspiracy with Perring and Wilson to steal the mine? It appeared that Wilson and Hamilton were members of the British Foreign

Service and stationed in Quito during that period of time. What was evident was that Mejia had been cheated out of receiving any funds from the transactions.

Barrana had been clever and sanitized his trail with half-truths, denials and giving Mejia false leads to the people that owed him 300 gold pesos. To the very end Mejia was unsuspecting of Barrana, for when he wrote his last will and testament he issued the following proclamation:

"Senor Adam Ricalde and Senor Raphael Cardenas gave the product 300 gold pesos and this money has not been given to me yet. I command my family to get this money and place it among my wealth." (Who these men were, or even if they actually existed, was not known as this was the only time their names appear in Mejia's documents.)

The twisting trail of deceit continued from documents recorded in Mejia's diary that led from the jungles of South America, to metallurgical laboratories in London and Paris, and back again. The British, the Colombians, and even the Catholic Church had explored the area searching for Mejia's treasure. As the web of intrigues ended, an interesting footnote was found on the inside spine of the back cover. It simply read. *Mr. H.L. Holloway, Mining and Metallurgical Club—3 London Wall Building—London EC2, England—Savoy Hotel, Quito.* That left us curious as to what role this British based company played in the unfolding drama.

It was late in the evening, when the maid brought in two steaming cups of strong Ecuadorian coffee; she had a difficult time locating a clean place to put them on the cluttered desk. Papers, books, maps and scribbled notes were lying everywhere. As the young Indian girl excused her self for the evening, I turned back to the family crest on the wall and asked Adriano about it. He proudly stood before the crest and told me his heritage was the Count of Vintimilla, and Marquis of Iberia, and the Kings of Italy. He was proud of his family's heritage and how his father, as a young man, attended the family's reunions in Italy. As time passed, although still noble with his royal roots, his father no longer allowed it to dominate or control his family's direction in life. He only wanted his two sons and daughter to have a good education, an honest scale of values and to treat everyone with equal respect. I realized how fortunate I was to have a friendship with this man.

Knowing that he was only two months from graduation made me appreciate the time and effort he devoted to the Mejia project. While he studied I wandered down the ancient city's main section of narrow sidewalks and rugged cobblestone streets. In the distance I saw hundreds of Panama hats that looked like a magical city of toadstools drying in the morning sun. Like the hats tediously woven knot by knot, most everything else in this quaint city was shaped by hand. Silver and gold filigree jewelry, rustic and modern designed ceramic pottery, embroidered shawls and skirts and delicately fashioned marble and woodcarvings; they all contributed to Cuenca being the artisan center of Ecuador.

Crossing the street in front of the post office the sweet smell of fresh baked bread filled my nostrils from the little bakery on the corner. It was here that my compadre from Yuma used to hang out every chance he got. The doughnuts were as hard as rocks but Bob said, "The terrific aroma fools the jaws and the digestive track." My thoughts were with Bob this day. I missed his humor and devil-may-care attitude and the camaraderie we shared during those long rainy days in this city. I hoped that everything was going well with him and his family. The loud sound of a car horn jolted me back to the moment. It was Adriano. I jumped in and we were headed to the Officina de Reclamation to solve another piece of the Mejia puzzle.

We entered a dilapidated building on Sucre Street that was filled with old typewriters and underpaid civil workers. Adriano was his usual self, dressed in dark three-piece suit, the formal antithesis to my casual windbreaker, turtleneck and Levis. Adriano asked for a map of the Sucua region in the Oriente and after a brief wait, the clerk returned with one that highlighted the area we were interested in. We were experiencing a bit of luck as the army was building a road in the area and had just completed aerial mapping of the region. It was synchronistic because our hand drawn duplicate of the diary map matched up perfectly with theirs. Anticipation built as we traced the outline of the Tutanangosa River, until it came to the Santa Rosa River near the headwaters up stream. It was unbelievable that after eighty-seven years the rivers had retained their original identity. It all looked so easy!

We were scheduled to meet with Alfredo thiat evening and organize our exploratory trip into the Macas Forest. This was the moment I had been waiting for. Alfredo arrived in good spirits and after the customary salutation of handshakes and bows we settled into the study for an evening of planning. Alfredo said he knew the lay of the land and the attitude of the Jivaro Indians that lived in the area. He reiterated that any expedition searching for Mejia's treasure would have to enter through the small Jivaro settlement of Sucua in order explore the Tutanangosa River. He warned that the Jivaro did not trust outsiders and they barely tolerated the Spanish Missionaries in the area.

We talked for about an hour and then settled in on a plan that appeared relatively simple. We would send in an advanced scouting party to push up water, along the north bank of the Tutanangosa. Alfredo would journey with a young Indian named Oriolfo who had worked for the Vintimilla's for three years. They would be our eyes and ears and under instruction not to discuss emeralds or Raphael Mejia. Instead they would establish a gold washing operation, taking in bateas and a seven-day supply of food.

Mejia had designed a similar ploy in a letter to a friend in 1908, when he wrote, *"In order to avoid alarm, we can say that we will make the partnership in order to get rubber and quinina and not emeralds."*

We set our target date for the exploratory trip to begin in one week.

I went to my room and made an audiotape for Larry Webb, owner of the Capitol Detective Agency in Phoenix. I filled him in on our research and investigation to date and accepted his offer to help arrange for heavy weapons. I was up front with him concerning the Ecuadorian laws on weaponry and advised him that whoever was selected, as the courier must be made aware of the dire consequences they could face if caught in Ecuadorian customs with the guns and ammo. The person would have to be fearless, yet calm and confident, if they were to succeed. My shopping list included an automatic shotgun, cleaning kit and two hundred rounds of 00- load ammunition, one carbine rifle with cleaning kit and two hundred rounds of ammo and a 357-magnum pistol with two hundred rounds. I also placed an order for three unique twenty-two rifles (AR-7's) and three hundred rounds of hollow point ammunition.

Afterwards I took a walk to clear my mind. My route took me by the movie house and I decided to check out what was playing. Needing an escape from the mental gymnastics of Mejia's last will and testament required a good American film, an action picture. To my dismay the film was an Italian melodrama with Spanish sub-titles and the thought of another two hours of translating left me weary and bored, so I made my way toward the park.

The City of Cuenca could easily qualify as one of the most picturesque cities in all of South America. The old Inca outpost that the Spaniards had conquered was a modern day blend of Spanish and French influence. In the mid 1800s, the old wealth of Ecuador claimed the city and moved onto large estates with showy palatial houses. Their heavy French provincial furniture and famous pianos were brought over the mountains on the backs of Indians from the port city of Guayaquil. It was the perfect place to escape the heat and grime of Guayaquil and the political ramifications of Quito and it was an ideal family environment. It had a moderate climate, schools, churches, parks and a prestigious university. But most importantly for me it was also the gateway into the Oriente.

Cuenca contained everything that you could possibly ask for unless you were an outsider and a bachelor who enjoyed good food, entertainment and female companionship. When the sun went down I only had a choice of two forms of entertainment: a movie or sitting in the park staring at the moon. There were two theaters in the city and each had only one showing per evening at 7:00 P.M. The American films brought in were at least a year or two old and lacked in quality, but they were in good condition compared to the French and Italian choices. There were no free standing restaurants in the city. Dining out was narrowed down to two choices, either the Hotel Cuenca, which I knew too well, or the Hotel Crespo that was a charming, old rickety building, perched on the side of the Rio Tomebamba. And to make things even worse, the city shut down at 10:00 p.m. Needless to say my evenings in Cuenca were lonely!

Entering the quaint park located in the heart of the old city was truly a unique experience for a foreigner, especially one that wore boots. Immediately six young boys between the ages of six and sixteen surrounded me with shoeshine kits and begged for my attention. I usually selected the youngest and propped my size ten's on his small wooden box that barely covered the length of my boot. The shine cost two cents, which was a real bargain, as they would always pour their heart and soul into their work. After they earned their wage and tip, the kids hung around to ask questions about that mystical land to the north known as Los Estados Unidos.

The more questions answered, the more they asked and pretty soon, the group of six would swell to twenty and I found myself back into the mental gymnastics of translation, which was what I was trying to escape. The light of the evening sun shadowed by the buildings around the square and the little group around me thinned out. Soon I was alone with nothing to do except get lost in my own thoughts, which was the other thing I was trying to escape.

At twilight people came in masses and the park filled rather quickly. At dusk the ritual of the promenade began on cue as it had for the last three hundred years. Families with their teenage daughters in tow glided with pride around the square. It was a time of family bonding and of showing off of the young Cuencano virgins being led around the square to the musical score of John Phillip Sousa. The scene was played out in every small village in Ecuador and was a big hit among the young studs of the village, including this stud from the United States.

Oh, those young puritan girls were downright playful. Flirting with their eyes and their smiles while waltzing by. Some of those smiles could leave you mesmerized for a week. The girls were safe and they knew it, like the Playboy clubs in the States where you could look but couldn't touch. The girls toyed with male hormones and the parents acted as if they never had a clue as to what these young ladies were up to. Pops was too busy checking out his competition while mom glared menacingly at all of us young guns. It was obvious who ruled the households in Cuenca.

When the music stopped and the street lamps flickered it was time to go home with the ones that brought you and there was no exception to this ironclad rule. The music ended and soon the park was deserted with the exception of a few packs of teenage boys huddled together discussing this girl or that girl. Soon, they too were gone and I sat there alone with my eyes closed, listening to hollow footsteps on the cobblestone streets.

My loneliness was interrupted by my shirtsleeve being tugged for attention. I opened my eyes and saw this old wrinkled woman in rags with her outstretched hand begging for a few sucres. I could never say no to the young or the old who were less fortunate than me. They were everywhere, even in Cuenca. For the old, my charity would buy them a few more worry free days of food and care. The young probably took the money home, if they had a home, to share with their family. I placed several coins in her outstretched hand and smiled up at her.

No, I couldn't say no, even when I was growing up on the Apache reservation; I shared my lunch with the Indian kids who had no food and my toy's with those that had none. It was the Christian thing to do my Grandmother always said, and she was a practicing evangelical. Although I had abandoned religion when I left home, in my heart I always knew she was right, that this was the way you treated fellow humans. Glancing at the clock on the wall of the bank, I realized that it was time to return to the fortress before those massive oak doors would be closed and locked for the night. I headed for Calle Borrero Street as the city began to slumber.

My restlessness was overpowering. It was time for a change of scenery so I informed Adriano that I was going to Guayaquil to check on the availability of helicopters. He understood. Adriano knew that there was nothing I could do in Cuenca except wait. The waiting game was the hardest for me, even if it was just waiting for a shipment of weapons.

The Hotel Humboldt hadn't changed much. The pool still looked like an algae pond and the smell from the restaurant upstairs still curbed one's appetite. I checked the Universio newspaper and the phone book for a helicopter company and made some notes. I couldn't wait for the Guayas Bay Club to open. I was looking forward to seeing Barbara again and devouring a plate of her club's signature giant prawns. But, that was hours away so I decided to go downstairs to the bar and kill some time and have a couple of beers. The bar attracted people off the street and was a magnet for soldier of fortune types, explorers, sea captains and tourists.

It was good to see Jose, the barkeep once again. The first drink was on the house and we reminisced about old times and updated him on Bob and our gold mining venture that was a bust. We were laughing about Washington's English ability when a grizzly old man sat down, ordered a beer and joined our conversation.

He introduced himself as Ed Walker from El Paso, Texas and claimed he had been in Ecuador for ten years, "Doing a little prospecting," as he put it. Jose left us to serve an older couple and I asked Ed what he was doing these days and for the next ten minutes he riddled around my question like a man on the lam.

"Not too bright, Lee," I chastised myself for asking such a stupid question to a total stranger, especially one suffering from El Dorado syndrome. The syndrome was like a mental illness that thrived on paranoia and the idea that someone was out to steal your gold mining claim, poke or the treasure you were seeking. Therefore, you never told the truth to any stranger especially another gold hunter, and never where you've come from or where you're going.

Everyone in exploration seemed infected with this strange illness. We went to great lengths to sanitize our truth and Ed had perfected the art of confusion and denial as an answer. He listed a dozen areas trying to throw me off his tracks. When he finished I knew it was my turn to creatively avoid answering. "What are you doing here? You don't look like the Galapagos tourist type," he sneered.

My syndrome kicked in hard and heavy and I created the most poetic story of photography and jungle wildlife that I could possibly imagine. I told him I was hoping to get into the Oriente by helicopter to do some serious photo work, before various species vanished at the hands kill-happy poachers. Ed blinked and ordered us another round.

"Let me give you some advice, son," he said dryly. "First of all don't go into the Oriente unless you absolutely have to. If you do wind up there, don't believe anything you hear and only half of what you see and get out as soon as you can. Second, never, and I mean never, spend the night on the east bank of the Rio Napo. That's Auca territory. They kill Indians and whites alike just for the hell of it.

"In 1937, the first Shell Oil Company geologists went in. They got a huge oil concession and they been under attack ever since. The Auca have killed missionaries, oil company workers, women, kids, and anyone who happens to venture into their territory. Hell they even kill birds! There's a story about a group of Spaniards looking for exotic birds, you know, parrots, things like that. They captured quite a few and put them in wooden cages. The Spaniards got attacked and those that survived had to high tail it out. They left the captured birds behind and when they returned with reinforcements, they found bodies hacked to pieces and do you know what these bastards did to the birds?" he asked angrily. Before I could reply he answered in graphic detail. "They tortured them by pulling out their feathers then sliced their little throats and let them slowly bleed to death." He stopped to light a cigarette.

"The Auca kill with long spears made from chonta wood. Both ends of the spear are pointed and one side has barbs cut in it. They tell me it is ornamented with colored feathers. They use blowpipes and arrows, too. The arrows are poisoned but the spears aren't. Curare dipped arrows are too quick for their crazed entertainment. They prefer to watch their prey ooze blood and suffer."

The lure of danger and the mystery of the unknown drew me in. I was so engrossed and drawn into Ed's graphic war stories that my planned dinner at the Bay Club faded from thought. We ate dinner upstairs and talked into the night. He'd studied everything he could get his hands on about the Aushiris Indians, also known as the Auca, which meant naked or savage. Slinging back a whiskey Ed called it a night and reiterated, "Stay away from the Napo, if you want to get back to Arizona." I nodded an affirmative.

That night I tossed and turned trying to shut off my mind and get some much-needed sleep, which didn't want to come. I went out onto the terrace, gazed up at the full moon and watched the ships turn on anchor, reversing directions in the timeless tide of the Guayas River. The serenity of the evening was as confusing as this contradictory land. Just a few hundred miles east the Auca were probably gazing at that bright orb planning their next killing spree. Then I thought about Mejia and why I was here and the pieces to the Mejia puzzle. Deep inside I knew that his clues wrapped in an intangible riddle would not be easily revealed.

CHAPTER 4

MEJIA'S RIDDLE

The Rio Santa Rosa—AKA—The Rio Cunguinza

The last will and testament of Raphael Mejia was recorded in the presence of an official of the court on December 14, 1887, some six years after he discovered the emeralds. The document was prepared in the Ecuadorian city of Tulcan on the border of Colombia and was in accordance with the law to insure the family's legal inheritance of his estate, holdings and claims. Since he considered the emeralds part of his property, Mejia was required by Ecuadorian law to give a detailed location where the stones could be found. Mejia, wary because of betrayal and threats, knew his enemies would take advantage of this public record. In order to protect the location of his treasure from claim jumpers, he contrived a riddle...the final clue leading to the resting place of the green fire. The clever puzzle baffled searchers, officials, and his family for 81 years. Now it was our turn to attempt to crack his code.

• • •

"I declare that in the womb of the Tutanangosa Jivaro tribe, Forest of Macas, I leave a very valuable stone that weighs tres arrobas..."

The opening statement of his last will and testament pinpointed the general area and did so without concern because the womb was a big, nasty place to begin any search. The Rio Tutanangosa was an artery flowing through the heart of the Macas Forest, the home of the Jivaro Indians. The headwaters of this serpentine river formed at an elevation of three thousand meters in a rain forest that can only be described as overwhelming. The jungle was triple-canopied, always wet, and the home of countless predators who were in perpetual search of food.

This was the collision zone for the warm, western drifting cloud cover of the Amazon basin and the cold eastern flowing cloud mass of the mighty Andes. When they collided, which was often, the Tutanangosa became a down hill torrent, rampant with a destructive force that could devastate anything and everything in its path. During the rainy season the river thundered with power. But even in the dry season it was formidable, flowing from west to east, eventually rolling alongside the small settlement of Sucua before it finally plunged into its muddy big brother, the Rio Upano. The Tutanangosa contained gold, boas and isolated emerald outcroppings. It was a terrible place to begin an expedition, but it was our first clue to solve the riddle.

Mejia declared that he found the emeralds in the womb of the Tutanangosa forest and evidence supported his claims and we believed what his will stated. We were certain that the first clue was solid, but then came the second clue, which split our ranks.

"This mineral stone is in the Rio Santa Rosa. Taking to the left...Oriente side, where there exists a landslide, from this landslide to the foot, which is about 400 meters in decent, a large stone open in the middle. In this, the deposit covered with other stones. In the front, runs the water of the same river about media quadra (50 meters)."

The Santa Rosa River was named after a Catholic saint by the Colombians. Santa Rosa was the first saint of the New World and the patroness of South America. Isabel de Santa Maria de Flore born in Lima, Peru, of Spanish parents took the name Rosa at confirmation. As a young girl Isabella experienced mystical gifts and visions of such an extraordinary nature that Church officials declared them supernatural and she was canonized in 1671.

According to the Army map it was the ninth river upstream from the settlement of Sucua. Located on the north bank of the Tutanangosa River, the Santa Rosa River flowed between the Rio Curinza and beneath the Rio Ojal, which was the final river that appeared on the map. Surrounded by other rivers bearing Jivaro names the Christian-named river was an anomaly that made Alfredo challenge the Colombian's writings. He felt Mejia was too clever and would never leave such easy clues.

Alfredo possessed an old map drawn by one of the Spanish settlers in Sucua. Surprisingly, the map indicated the cascarilla trail from Quito into the Macas Forest ended on the south bank of the Tutanangosa alongside a river named the Cunguinza. Alfredo believed that the cascarillas had renamed this river, the Santa Rosa, for easier identification and pronunciation because the Jivaro name for the river would have confused the workers. He argued that the men were all Spanish, brought up with almost zealot religious convictions, and they would naturally prefer a name of a saint, rather than something named by a savage.

His rationalization was sound because the old cascarillas trail was parallel to the Cunguinza, which flowed south to north, until it flowed into the Tutanangosa. His strongest argument was that there were no bridges across the Tutanangosa River. The cascarillas would not have chanced placing their base camp on the north bank, the opposite side of a completely unpredictable river like the Tutanangosa. Alfredo wanted to change our original plans of searching the north bank and move the entire expedition to the south bank. His argument made a lot of sense and brought back memories of lessons learned from Little Hell.

Adriano countered. His argument was purely pragmatic, which reflected his major in college. His law background dealt with facts and not supposition. The Army map had clearly identified the Rio Santa Rosa on the north bank of the Tutanangosa, one river downstream from its headwaters. If the cascarillas or Mejia had changed the Cunguinza name to the Santa Rosa, then it would have appeared as such on the army map, but it did not. He also argued that weather patterns changed through time and eighty-seven years ago the weather conditions might have been different. The Tutanangosa might have been a small stream then. Alfredo had no defense for these points. Then Adriano took his argument one step farther by quoting from Mejia.

"Crossing the first river up stream," Adriano argued could only mean one river. The Santa Rosa! It had only one river upstream and this river was named the Ojal. The Cunguinza had no rivers upstream according to the map and it was on the opposite bank therefore, in Adriano's pragmatic reasoning, we must go up the north bank, all the way to the Rio Ojal, the headwaters of the Tutanangosa.

"The color of the stone is white and in the womb green, the shape is large. And in the place where it exists, is a clean rock and at the foot is the indicated stone."

We surmised that Mejia was describing the white matrix containing the emeralds and a landslide containing the emerald outcropping. "*...And at the foot is the indicated stone."* The indicated stone would be the one, open in the middle where he hid the tres arrobas after dulling his machete and breaking the stock of his shotgun separating the stones from matrix. Next was the puzzling riddle.

"Like making a relation, the entrance is by the cascabel and it takes crossing the first river upstream."

And the legend continued. This was Mejia's final clue to the location of the emeralds that he had hidden in strange words. *"Like making a relation...*" left us shaking our heads. Did he mean something like a sexual union or going into something or did he mean like bonding, creating a relationship? We couldn't even begin to hammer down a definitive answer to this phrase. "*The entrance*" was also confusing. The entrance to what? But the ball buster was Mejia's reference to the word "*cascabel*". What was the wily Colombian saying? We were stumped over the word cascabel. It had no logical translation for our application.

There were only three Spanish translations of cascabel; a bell, the knob on the end of a cannon and the name of a snake found in the Oriente that, like a rattlesnake, emits a rattling sound of warning before it strikes. Growing up on the San Carlos Indian Reservation I had many close calls with rattlers as we called them so I related to the cascabel snake in the Oriente. But how the cascabel fit into Mejia's puzzle was a mystery. We decided it had to be a natural landmark because he was creating a visual trail for his family to follow.

Mejia used cryptic definitions and puzzling words in his public declaration to hide the location of his lost treasure. I couldn't help but wonder why he didn't he take his family aside after the notary left and just tell them what these strange phrases really meant? After all he was adamant when he said, "*I command to my relatives to make the expedition and put this among my wealth.*"

Adriano translated my question to Alfredo who replied in a rapid outburst of Spanish that seemed to last ten minutes. "It seems like he did, Lee. The relative of Mejia, the woman in the Oriente with the two kids that was befriended by Alfredo said the family was instructed to remember the keys to the location of the emeralds. But their search proved unsuccessful. Perhaps, through time they forgot some of the important clues or became confused. The family never found the treasure."

The maid brought us coffee and rock hard doughnuts to sustain us through the debate that continued into the late hours of the evening and was only adjourned because of mental exhaustion. We concluded that to solve the riddle required men in the field and we would stay with our original plan to send Alfredo and Oriolfo up the north bank on a seven-day fact finding expedition, which was only two weeks away.

I asked Adriano for a map of Ecuador and he went into the study and produced one from his desk. He unfolded the large map with a hint of curiosity. I told him I was looking for the Rio Napo, which he quickly pointed out. I found the Tutanangosa and took a deep breath. The differences between these two rivers were measurable. Now, at least I knew, we wouldn't be camped on the doorstep of the Aucas.

Raphael Mejia's last will and testament was dated December 14, 1887. The first two pages dealt with the last will and testament riddle. The next four pages were Mejia's instructions to his family to pay off debts and give a percentage of the wealth created from the emerald find to various churches. Pages seven through forty-two were old letters to and from various friends of Mejia. How the family wound up with the Barrana letter was a mystery. Pages forty-two through one hundred and six were a history of the progress made in the Oriente that covered everything from rubber gathering to bridge building. The information found in these sixty-eight pages, was totally insignificant to the emerald story but an interesting historical document. The challenge was to unravel the mysterious location, which were hidden in the first two pages.

Mejia's clues gave me a nagging headache. Every time I tried to decipher his verbiage I created more questions. Not only was each segment difficult to wrap my mind around but also the overall riddle was beyond comprehension. The Macas forest was a big place to explore. We had eliminated the derumbe (landslide) clue because the jungle would have devoured any exposed rockslides over the years. The Inca built an elaborate road system throughout the Oriente that they were forced to supply with full time maintenance crews just to keep the roads from being swallowed up by the prolific jungle growth. Landscapes changed quickly down here.

The most puzzling aspect to Mejia's claim was, *"the entrance is by the cascabel."* When he used the word entrance, was he talking about a tunnel or a cave entrance? Or was he describing an area, like a meadow or perhaps a small arroyo or ravine that led toward the jagged rock? As I banged my head against Mejia's double innuendoes, I suddenly heard the gate downstairs open and close and footsteps up the stairs. It was Adriano. He joined me in the study and asked what part of the riddle was causing the anguish written on my face.

We retraced our original steps and re-read the old letters looking for something more tangible that referred to the entrance. Then Adriano breathed a deep sigh of relief. "Here it is Lee. A letter from Mejia to a man named Canal."

"If you try to look for this wealth, I will do as follows. I will send four workers under the direction of Soto, an honest man, the type of man, I will always trust. And I will draw you a map and give you the right way to get to the emeralds. I will also tell you the place where there exists a socovom."

Adriano paused in his reading and while I waited in anticipation, he reached for a large Spanish dictionary. "This word socovom has no meaning. Oh, I see," he muttered to himself. "The word is misspelled; Mejia used a v instead of a b. It should be spelled socobom, which means cave, cavern or tunnel." Adriano then continued reading.

"...And also a mill and ruins of houses...These places were found by me, before I found the emeralds, we didn't pay much attention to these ruins, because for us the quinina were the best treasure for us, in those happy times."

Now we were really confused. After dissecting the letter piece by piece, it became apparent that the socobom had nothing to do with Mejia's mine, because he said, "that he found it before he found the emeralds." The part about the mill and ruins of houses intrigued us, but we concluded that it must have been a previous Spanish mining operation.

So we went back to square one and concluded that the word "entrance" could refer to a litany of explanations. We placed it in the category of unknowns, along with "cascabel" and "like making a relation". But, the real disturbing aspect of this letter was when Mejia said. *"I will also draw you a map and give you the right way to get to the emeralds."* This led me to believe that Mejia's hand drawn map in his last will and testamentand was a ruse. Was I over reacting, or was Raphael Mejia dragging another hare across his trail?

CHAPTER 5

PADRE CRESPI

Lee Elders and Padre Crespi
Lee holding two bronze plaques, one clearly depicting an Assyrian winged bull. On the counter the eagle headed genie as seen in the temple of Ashur-Nasir-Apal II in Mesopotamia.

Adriano had a surprise waiting for me. I retrieved my camera and tape recorder from my room and we drove to the outskirts of the city where his mystery unfolded. He parked on the corner of Padre Aguirre and Vega Munoz Street and we walked into the Salesian School behind the Maria Auxiliadora Church. He asked for Padre Crespi. A jovial old monk with a flowing beard who was Cuenca's most famous eccentric and revered personality soon greeted us. About forty years ago Crespi, as a young priest, caused international ripples and ridicule when he claimed to have found Mesopotamian artifacts that proved that ancient civilizations from the Middle

East had established trade routes with societies near Cuenca.

Carlos Crespi was born on May 29th, 1891 near the northern Italian city of Milan, where, as a youth he lived a comfortable life with his family. After he graduated from Milan University with a master's degree in anthropology, he mysteriously gave up everything to join the Salesian Fathers whose order was based on voluntary poverty. Over fifty-five years ago the missionary order had sent Father Crespi to South America and for that length of time he had been dedicated to his faith, sleeping on the floors of native huts, eating simple foods and doing without modern day comforts. He cared for the people, giving mass, baptisms and listening to the poor peoples hopes, dreams and despair.

Entering the large, dark room that only Father Crespi had keys to, we paused as he fumbled with the light switch. Then, with a spark, the room was bathed in a soft, yellowish light that brought the past to life. Scattered haphazardly throughout the dusty, dank smelling time vault, were thousands of artifacts that seemed frozen in time, each with a story that only *it* was capable of telling.

Lined up against two walls were countless metal and ceramic plaques, Tiki Gods, Inca pottery, and lying on a glass case, was a two-foot alligator possibly fashioned from gold? Next to this were two bronze plaques, one clearly depicting an Assyrian winged bull; the other appeared to be Inca. Propped on the counter gathering dust was a splendid replica of the eagle headed genie as seen in the temple of Ashur-Nasir-Apal II in Mesopotamia. This plaque weighed close to ten pounds and, if authentic, would have dated from the middle of the 8th century BC.

Almost apologetically, Crespi explained that the original museum housing the objects had burned down in the 1950's, and many pieces had been destroyed in the fire that engulfed the building. But apparently thousands had been saved and here they were stacked endlessly on top and around each other.

Crespi took great pride in the massive collection that he said came from Inca tombs, ruins, and vast caves found deep within the jungles of Ecuador. He agreed to pose for photographs with select pieces. Adriano snapped a photograph of the priest holding the golden alligator while I held two heavy bronze plaques, one whose style was clearly Inca and the other breathtaking plaque clearly representing the Assyrian culture. We filmed solid gold, Inca facemasks, breastplates, shoes, necklaces scepters and shields. Then Adriano filmed Crespi and me with two gold Inca masks that we placed in front of our faces for the sake of the photo. The dilapidated old room held a fortune in silver, gold, Pre-Colombian and objects-de-art whose provenance was mysterious to say the least.

Adriano cautioned me to conserve film as the priest disappeared into a back room. Soon he returned with a chunk of white matrix the size of a soccer ball. Close examination revealed approximately twenty emeralds of all shapes, sizes and different shades of green color. Immediately, a quote from Mejia's last will and testament came to mind. *"The color of the stone is white and in the womb green."*

"Where did this come from?" I asked, focusing my camera.

"Rio Tutanangosa," Crespi replied in a high pitch voice.

"My God," I mumbled.

Adriano was grinning from ear to ear; I could only nod my head and smile. My friend had done his homework, when he sent me the letter saying that, "he had uncovered evidence that supports Mejia's claims." Padre Crespi had found a cluster of emeralds in white matrix from the river that carried the same name as the forest.

We had hard evidence to substantiate Mejia's claims beginning with a military aerial map identifying the same rivers and now a cluster of emeralds in a host rock found in the Tutanangosa River that Mejia wrote about some eighty-seven years ago. And our star witness was a dedicated man of the cloth. As my friend Bob Olson often said, "It doesn't get any better than this".

It was equally significant that he had been very active in exploration of a country that even in 1968 was still fifty percent unexplored. In 1923, Crespi led an expedition up the Rio Tutanangosa, after Indians in the area reported that heavy rains had dislodged green stones and much gold in the river. He never once said that he was searching for emeralds, but he did say, "Indians found thirty-five million Sucres in gold in the Rio Tutanangosa after a big storm...and the emerald rock." It matched the discription of the host rock that had been found in Mejia's river.

Crespi's expedition up the Rio Tutanangosa was wrought with danger. On the second day they lost one of their workers to the deadly equis snake. Crespi said, "It had been raining all day and finally when it had stopped, the workers were chopping palm leaves and harvesting other vegetation to build a fire. We were trying to dry out and stay warm. I asked one of the cargo carriers to go and get some matches out of my backpack, and when he reached inside for the matches, the snake bit him on the hand. The Jivaro worker died a horrible death," he said sadly. "While I was administering last rites, blood started popping out of the pores of his skin". The equis snake gets its name from the letter x and the Indians in the Macas forest named it because of the x patterns that run the length of the body.

On the third day, they were chopping a trail through a large strand of grass, when they came across the bones of what he described as a young boy who had either been bitten by a snake or had fallen asleep in the path of soldier ants, which Crespi said was the most dangerous of all the predators in the Oriente. "They will attack man or anything that happens to be in their four column wide path as they march through the jungle, in a constant search for food. And their bite is very painful."

Crespi was bent and aged and wearing a frock that was old and fading from black to gray. His scraggy beard covered many of the hard lines etched on his face from years of hardships and sacrifice brought on by his servitude to his Church Order. He was 77 years old and normally displayed high energy and enthusiasm, but on this day he appeared tired and distant as he placed the long key into the rusted lock and opened the old wooden door that led to his private treasure.

Crespi jiggled the loose wiring and the lights flickered, casting an eerie glow in the dusty, dank smelling time capsule. The room was still crammed full of unimaginable objects that looked like ancient and prehistoric artifacts. Admiring dozens of chair size urns with distinctive geometric designs, which appeared to be authentic pre-Colombian pieces, Adriano asked the Priest how and why he had amassed such an enormous quantity of objects. Crespi replied, "Oh, this is nothing compared to the treasures that are still out there in the vast caves and tunnels that have not been discovered in the Oriente."

Crespi was making reference to the places where they found artifacts. Hidden in the immense jungle according to Crespi were large pyramids, deserted cities, enormous sacred tunnels and interlocking caves. He had been told that the cities shined with a mysterious bluish light when the sun goes down. The tunnels were large enough for ten men to stand side by side and have cut-stone entrances and walls as smooth as glass. It was a fantastic story that he took great pleasure in telling and that led many to dismiss them as a figment of the old cura's imagination. But, when you stood in his presence and examined the private treasure he controlled, my mind accepted his fabulous tales as authentic. He was adamant the Indians were the key to this wealth. "Without their trust and friendship the Church would never possess this enormous wealth," he said.

I wandered aimlessly through ancient stone sculptures and carvings while pausing briefly to examine a gold tiara with linear writing. I noticed that Adriano was having an animated conversation with Padre Crespi and motioned for me to join them. "I asked, if Padre Crespi would show us his private room today and he said that he could not because his time is limited and we must hurry to the porch," Adriano said. My God, I thought to myself, if this room is open to the public and the other room is private, what in the world is in the second room?

We entered the porch area off to the side of the public treasure room we had just visited. Crespi had noticed my interest in his expedition up the Rio Tutanangosa, especially the part where the worker had been killed by the Equis snake. He steered us towards a shelf that held numerous bottles and jars that contained well preserved snakes and spiders in alcohol. "Equis," he said while pointing to a jar holding a small brown and white snake that measured about ten inches. If the Jivaros are afraid of this little guy, he must be one mean son-of-a-bitch.

Next, he pointed out a beautiful red snake in a large bottle and next to it a deep green snake with an ugly head. Both were poisonous, he said, and then he chuckled and shared a story about the time he was on an expedition near Mendez. They had chopped their way through some dense jungle and had entered a large clearing when they spotted the green Mamba looking snake about the same time the snake spotted them. The green snake flung its body around, headed straight toward them, riding with its head about four feet above the ground and traveling the thirty yards in a quick burst of speed. Crespi's workers armed with machetes waited for the deadly intrusion and when it came in close they hacked it to death. "It is very brave and will persecute a man," he laughed showing stained teeth. Crespi understood English, but preferred to speak in Spanish or Italian. He added, "This snake is not poisonous, but cunning and powerful." He pointed to a twenty-five foot boa skin draped across an entire wall.

The old bearded priest patted my arm, "Be careful near the Rio Tutanangosa and the Rio Upano. The Boa lives in those rivers and will grab you and pull you underwater. They drown you and then will take you to an underwater lair and feed off you for weeks." His graphic translation lingered with me; his dire warning was heeded. About the time we were comfortable with poisonous snakes staring passively from their final resting places, Father Crespi waved us over to a row of glass jars containing red and brown spiders that, according to Crespi, were worse than the Equis. One jar contained a spider the size of a coffee can lid. He held my sleeve and stared into my eyes. "They are very poisonous. They are found in the trees and heavy brush. You must be very, very careful when you take an expedition into the Oriente. There are many dangers lurking there."

His intense stare and thorough warnings gave me the distinct feeling that the old man was aware of the emerald project and was testing me to see if my heart was really in it. If a person could make it through his chamber of horrors and still have the courage, or ignorance, to continue on, then one would have to assume they had passed the test. I wondered if he was trying to scare me off or if, in some strange way, he was trying to protect me by informing me? He was a cagy old cura and you couldn't be certain what was going on in his mind.

There were rumors that the Salesian school behind the Maria Auxiliadora housed two more rooms, one containing precious gems stripped from stone statues, idols and artifacts. The other room, off limits to the general public was rumored to be the stripping room storing discarded stone figures. Adriano, who had lived in Cuenca all of his life, had heard stories about these closely guarded rooms, but had never been able to gain entrance. "I have asked Father Crespi to show us these other treasures, but he has always been too tired or too busy. I guess you have to catch him in the right mood and the right frame of mind," he said as we stepped outside.

I remarked about how easy it would be for someone to kick in the flimsy wooden door to the treasure room and steal the priceless objects. Adriano explained that there was no security at the museum, but it didn't matter because the security was located within the police checkpoints ringing the city of Cuenca. All cars leaving the city were stopped and checked for contraband. During the day the Salesian school is home to hundreds of poor boys rescued from the streets. Their allegiance is to Padre Crespi and the other priests, who clothed, fed and educated them. The school had its own youthful security force of hundreds of eyes watching and reporting anything unusual to their guardians. It would be almost impossible for anyone to steal from Crespi and get away with it.

Crespi became tired and looked weary. He said it was time for us to leave. There was something about his eye contact that told me that our destiny with each other was just beginning.

• • •

We drove back to the Vintimilla residence in silence and all I could think of was the myriad of events that had occurred in the past week. We had successfully tracked the emeralds of Mejia through painstaking research that required countless man-hours of reading and translation. The time frame of events supported the claims found in his last will and testament. And topping everything off was Padre Crespi's 1923 expedition up the Rio Tutanangosa that had produced the white matrix rock that contained twenty-five emeralds. The white matrix rock matched Mejia's description perfectly. Crespi's discovery supported the area as an emerald producer.

CHAPTER 6

GREED

Mythological man-snake encrusted in emeralds and rubies Found by Padre Crespi in Oriente cave

Barbara Marsh was a magnate for soldiers of fortune and adventurers that frequented the upscale restaurant and casino that she managed in Guayaquil. The mysterious American from New York had lived in Ecuador for thirteen years and had seen and heard the stories of treasure hunters. When we first met she took me under her wing and shared her knowledge of the explorers that had walked a similar path to the one I was navigating.

Jim Bell, my gunrunner, cleared customs and safely stashed my new weapons and ammo in the hotel room. We drove to the Guayas Bay Club where Barbara joined us and ordered a round on the house to celebrate what she called, my return to "Condor country." She saw our puzzled looks and explained that Ecuador's national bird was the giant Condor and we could read into that little bit of trivia however we wanted. As the night wore on music played and wine flowed, we truly enjoyed each other's company and stories. It was one of those magical evenings that you wanted to last forever.

Jim was inebriated and leaned back, closing his eyes, a big smile on his rosy face. He looked like a sleeping Buddha for the rest of the evening. Barbara was curious as to what I was doing, so I told her about my planned expedition into the Oriente. I asked, if she had ever heard of the Mejia emeralds. She said she hadn't, but she was aware of the lost Connelly emerald find. For the next hour, while Jim slept in a sitting position at our table, she told a story about one of one of the most incredible lost treasures: the story of Stewart Connelly's harrowing adventure, his success in finding a rich emerald deposit in the Oriente, and how it was all documented in the archives of the Director of Mines in Quito. The record of his incredible story exists on a few yellowed and partially destroyed pages that the young man from northern Illinois scrawled in his personal journal to establish his legal claim to the mine.

Stewart Connelly enlisted in the U. S. Army and upon graduation from high school and served with the infantry during the World War II. After the war he stayed in Europe, finally settling in Madrid, where he spent his days in the Bibliotheca Nacional, studying the fascinating volumes on Pizarro's conquest of the Incas. Written by the Monk Sanchez, one of the friars who accompanied Pizarro's expedition recorded a fascinating account of the emeralds that were given to Pizarro as a token of friendship. The friar's records would be responsible in changing Connelly's life and leading him into the Republic of Ecuador.

Connelly journeyed into the port city of Guayaquil and eventually made his way north to the city of Quito. He located a small house on the outskirts of the city and with his meager funds, purchased a few books on Indian dialects and begins learning the strange languages while pouring over the old maps in government offices. Then, with his remaining funds he made a purchase, which to any reasonable man would signal a sign of insanity, but for Connelly, the bamboo flute was his passport to emeralds.

From his studies in Spain, Connelly had learned that for an outsider to enter the forbidden territory of certain savage Indian tribes was to invite certain death. But Connelly had a plan. In the event he was captured he would convince the Indians by blowing shrill notes on the flute and acting the part of a crazy man. He had learned from his research that some primitive Indian tribes had befriended and even venerated those they thought were insane and Connelly had decided to risk his life on this information.

Connelly accomplished his mission. According to the story he was captured and not killed by a primitive tribe because of his strange behavior. He escaped after being shown green stones by the tribe that had accepted him as a court jester. But he paid the piper. In his haste to escape he had left without his trusted compass and for weeks he wandered like Moses through the wilderness searching for a Promised Land called civilization. He ate roots, insects, anything to maintain his energy and strength. Finally, on the verge of starvation, a Josefina missionary found his body lying along the bank of the Napo River and rescued him. Tied around his neck was a bag of emeralds.

It had been almost a year since he left Quito armed only with a compass and a small bamboo flute. Now he was returning a hero with a sack of emeralds that many said were the finest gems seen outside of the Muzo mine in Colombia. Upon his return to Quito, Connelly could have sold his emeralds and retired to a life of luxury, but, unfortunately, news of the emerald strike soon leaked out. A small army of treasure hunters, mercenaries, and adventurers convinced Connelly that he should return with them to the emerald mine.

Connelly formed an expedition with about twenty courageous men armed with shotguns, pistols and carbines to make the dangerous journey beyond the Napo. With a half-dozen pack mules, ample food, and ammunition to last for several months,

Connelly and his band of adventurers left Quito and headed into the Oriente. Connelly believed that his heavily armed force would be capable of subduing any hostile Indians they might encounter, and therefore he would not need his magical flute. Weeks turned into months and months into years. Connelly and his private army were never seen again. The dense jungle had swallowed them up and to this day, their fate remains one of the great mysteries of Amazonian jungle lore.

"So, my blond haired friend, don't forget your bamboo flute," Barbara added with a hint of sorrow. She waved the waitress over to the table and ordered three large Ecuadorian coffees. Barbara continued as the piping hot coffees arrived. "Lee, why are you in Ecuador? What's your motivation? Where does your adventure end and greed begin?"

"Greed?" I blurted out. I wasn't prepared for the question. I had to think for a minute before answering. "I came to Ecuador because of money. I wanted to find that pot of gold in some river. I found my El Dorado, but then, my priorities changed during my second expedition to Little Hell. And when Bob nearly died on the trail... well, I could have sent him back with the Indians and gone on for the gold. But I couldn't so I went for help..."

Barbara interrupted. "So you defeated the greed for gold through your compassion for a friend? Is your emerald project based solely on adventure?" Barbara reached for her coffee and took a sip studying my reaction. I was debating the question in my Gemini mind. "What will you do if someone's greed surfaces in this project? Rumor has it that Connelly's final expedition never reached the Rio Napo. Greed among his team of mercenaries resulted in them killing each other."

Barbara was on to something. Although my mind was on overload between alcohol and her startling tale of Connelly and his demise, she had my full attention. I had faced betrayal but never the cancer named greed. It was true that gold fever had destroyed many men that searched for and found the tears of the sun. But gold was plentiful and the greed disease associated with it was only diagnosed as a fever. Emeralds were rare and their lure was a plague beyond comprehension. Pizarro, the Spanish Conquistador succumbed to their temptation and his avarice was apparent when he told Sanchez, the Monk. "...I want not seven, but seven hundred, yes, seven thousand..."

What would I do if greed became a part of the Mejia project? I carefully selected my words in reply to Barbara's question. "My associate, Adriano, and I have never discussed the money aspect in the Mejia project. Our mutual excitement and motivation is to break his code and the discovery. If I'm successful in this project then the money becomes a by-product of my quest for adventure. If greed rears its ugly head then I'll be forced to make a decision that I hope will be right, not only for myself, but others that I'm involved with." I paused long and hard in thought. "Where does adventure end and greed begin? That's an interesting question that psychologists would have a field day with. I personally believe that the answer lies within your true values."

"But, what if someone you are involved with succumbs to greed? Maybe even someone searching with you? What will you do?" she asked.

"I'd be forced to make one of three decisions. First, I could take a hike from the project, which is probably the worst of the options. I'd be giving in because of someone else's greed. The second would be to try and reason with the person and convince him why he needs a team effort rather than an individual effort. You know, the old adage that there is safety in numbers in the jungle and mountains might be a strong enough deterrent to make the person think about the consequences of eliminating his partners? My third option would be to hand him a one-way ticket back to where ever he came from."

"Did you ever see the old Bogart film, "The Treasure of Sierra Madre," she asked coyly. I nodded. "Then you remember that great scene when Bogart crosses the threshold of greed. We know it because his eyes start to betray him. And you know from those jerky eye movements that the man is suspicious that his partners are going to steal his share and maybe even kill him for the gold. Whether it's true or not doesn't matter to Bogey. He's a very dangerous man to be with and his will to reason deserted him along with his morality. Think about how you'd handle a Bogey, Lee." She knew I didn't want to go there, but continued anyway.

"Do you think Connelly made a mistake going back for more emeralds with a private army?" she asked while motioning for me to retreat to her private office in the rear of the restaurant. On the wall behind her desk is a huge map of Ecuador. As her fingers began tracing rivers in the Oriente, she paused while waiting for my answer.

"Yes, I do." I answered emphatically. I thought Connelly's sense of judgment was clouded by bravado of the herd mentality by a bunch of guys who were probably all down on their luck. It was Connelly's ego that betrayed his sense of judgment. The wild bunch that sought him out had nothing to lose but he had everything to lose. Shit, he was the first white man to discover an Inca emerald mine and he had proof. He could have sold the emeralds and retired, told his story to his grandkids, but instead he strapped on his backpack and challenges danger from the elements, the rivers, the tigers, the snakes, the savage Indians, and betrayal.

"Look, here is the river Shiripuno," she said, pointing to the map. "This is where Connelly first made contact with the Orijones. The Shiripuno River lies in the heart of Tribus Aucas. This is Auca Indian Territory! Connelly really pushed his luck by attempting to go back into that area with a group of misfits. Sometimes a bird in the hand is worth two in the bush, don't you think?"

For the next hour I listened to Barbara discuss the hopes and dreams of the many men, the explorers and adventurers, who had confided in her as they passed like ships in the night through the portals of the Guayas Bay Club, in search of fame and fortune. "Many failed and left the country too embarrassed to even say good-bye," she said sadly. "And then there are the others who never came back at all. Do you know about the reports of people literally disappearing into the Oriente, never to be seen or heard from again? Then there are those, those poor sob's that were physically and mentally broken by the Oriente." She offered a glass of cognac, which I declined, poured a round anyway and continued.

"Last year, I got attached to this wonderful man, in his late fifties, a handsome, ex-British paratrooper searching for the famous Valverde treasure. The Incas in the Llanganatis Mountains buried the legendary treasure? Ruminahui, an Incan Prince and favorite General of Atahualpa, was said to have buried a huge treasure of gold and jewels. I warned. I pleaded. He laughed it off and said that he'd probably seen worse. Well, to make a long story short, he left for the small village of Pillaro, near the Llanganatis Mountains in central Ecuador. "Here," she pointed to the map.

"A month passed, I figured William was a goner. Then one night he appeared in the restaurant and tapped me on the shoulder. I turned around, and believe me; I didn't recognize him he'd aged so much. When he said my name I recognized his voice. They'd finally reached the base of the Llanganatis, but had to turn back because they lost all their supplies and food while crossing some river. He said he wished he had followed my warning. Those ghostly mountains bring on a state of depression that consumes one's very soul. I hope I never experience that feeling again, as long as I live." She looked me square in the eye.

"He went back to England; a broken and sad man. I never saw or heard from him again. Lee, promise me two things before you leave tonight. Promise that you will always stop in and say goodbye, before you leave for the States. Last year, you and Bob left without letting me know that you going home and for a whole year, I thought your remains were east of Nabon somewhere and it really bothered me. You're a special person, a friend, so please check in every now and then? Let me know that you are alive and well. And never go into the Llanganatis. Okay?"

I nodded. But inside, I knew I didn't mean it. I had already made myself out to be a charlatan when it came to these things. I told the boys on the trail to Little Hell after they warned me about the Oriente that I had no desire to go there and three years later

I'm planning an expedition into the region they feared.

A ringing sensation in my head brought me out of a sound sleep. It was Adriano calling from Cuenca informing me that the boys, Alfredo and Oriolfo, had left for Sucua two days ago by DC-3 and should be headed up the Rio Tutanangosa as we spoke. Adriano had supplied them with money, a six-day supply of food, an AR-7, ammo, and two portable radios for barter. He felt they were confident and excited about the trip. His voice trailed off and the telephone cracked loudly in my ear. I asked if he could hear me. Hearing a weak, "Yes", I screamed even louder into the phone and woke Jim with a jolt. "I'll be in Thursday morning and will make arrangements from here for a bush pilot to fly us to Sucua to pick up the boys and bring them home. I'll see you Thursday morning," I yelled.

The travel situation in Ecuador was an adventure unto itself. The road from Guayaquil to Cuenca was still under construction and only the brave or poor dared to travel by car or bus. Locals half teased that there were two prerequisites for attempting such a trip: a death wish and a prayer rosary. But, if you thought you were safe in the friendly skies, you were wrong. One airline cargo carrier went out of business after only a few years because they lost all their aircraft in plane crashes. Not because of bad pilots, but bad weather.

The pilot's that flew in Ecuador were brave and a hardy bunch that were skilled in flying, but the weather problem was always in back of their minds. Flying from Guayaquil to Cuenca and into the Oriente, the pilot and crew might encounter three different seasons on one flight. Pilots had to be constantly alert; the weather could change dramatically in minutes and was the major contributor to so many fatal crashes.

I had my work cut out; I had to find an honest chopper pilot that would be willing to fly me into the Oriente. Wilbur Criswell was a twenty-eight year old British helicopter pilot who had been flying in Ecuador for eleven months. He flew choppers that sprayed the banana plantations with insecticide, protecting the 'yellow gold' that was the number one export for the Republic of Ecuador. We got along well and he was eager to assist me in my search for a helicopter that was capable of flying over rugged terrain. I had met him at the Humboldt bar when Bob and I were searching for dredge parts a year earlier and we stayed in touch. I rolled up to his office on the far side of the airport, hoping he would have some new information.

There were three helicopters in Guayaquil. Two of them belonged to the San Carlos Sugar Company and they were not for lease. The other chopper was a small Hiller, and it could be leased from his company. I explained that my destination was in the Oriente, a place named Sucua, and I needed to fly up a river whose headwaters were around nine thousand feet in elevation. I needed a helicopter that would carry a minimum of three people and a cargo load of at least two hundred and fifty pounds. Eyes wide he quickly ruled out his little Hiller and suggested a trip to Quito where I could talk with a company named Ecuavia Oriente.

Wilbur didn't mince words. He said it would be expensive, roughly twenty-five hundred dollars a trip and I'd have to pay up front, whether we made it to Sucua or not. If the chopper sat on the ground eight hours in Cuenca because the weather was bad, I still had to pay. He suggested I really think it through. The chopper only flew 80 miles per hour and I was looking at a fourteen to sixteen hour flight between Quito and Sucua. The problem with some of these choppers is that you can't fly as the crow flies. They just don't have the power to get over some of the colossal mountains.

Wilbur's sobering information left me with only one option, a plane. He gave me the name of Atessa Aviation and suggested that I call them to arrange a charter flight into the Oriente. He said to contact Jorque Merchand or Captain Drexel with Atessa, pilots of fixed wing aircraft at reasonable prices.

I grabbed a cab and headed for the Atessa Aviation office. The office manager listened to my frantic plea for a charter plane and placed a telephone call to Jorque Merchand who was in Cuenca.

After a brief discussion Merchand agreed to fly Adriano and me to Sucua in a Cessna 180 to pick up Oriolfo and Alfredo. His parting words hung in my mind like the clouds he was concerned about, "We have to leave early. Before those damn thermal clouds start rolling in.

The following day everything went according to plan. Jim left early on Peru's national airline, Aerolineas Peruanas, which originated in Lima, with a stop in Guayaquil before deadheading straight to Los Angeles. I was beginning to feel a twinge of home-sickness when I made my way back to the hotel room. Before long, the twinge I had contributed to nostalgia was now erupting into full-blown stomach cramps and diarrhea that had my head reeling and my legs running to the bathroom. I groaned while rummaging through my suitcase searching for a medical cure. My last day of freedom was spent in bed and in the bathroom. I couldn't wait until tomorrow to get the show on the road.

The next morning, I made my way to the SAN ticket counter. SAN was synonymous with National Air Service and was one of Ecuador's better airlines that flew into Cuenca, Ambato and Quito. My head still reeled from my bout with Atahualpa's revenge. The pretty girl at the ticket counter shook her head as I approached her for my ticket. God, I must look like hell, I thought as I leaned across the counter with sucres in hand.

"Sorry Senor, there are no seats available on this flight." she said, in a beautiful dialect of English-Spanish. "The tour busses have not arrived yet, Senor."

When the busses arrived, I recognized many faces from the restaurant the night before. The same people that I previously wanted no interaction with were now my sole priority. I tried everything to get on the plane to Cuenca. I offered double the price for a ticket, but to no avail. They came together and they would leave together. The ticket girl, feeling sorry for me, steered me towards TAME, the other airline, but it had no flights to Cuenca until noon and that would be three hours after the fact. I knew that I was going to miss my charter flight to Sucua and my people in the Oriente were going to be disappointed and confused and Adriano would be mad as hell.

CHAPTER 7

DEAD TIME

Lee and Adriano preparring for a charter flight to Sucua

The Cessna 180 left Cuenca at 10:00 A.M. with pilot Jorque Merchand at the controls and four replacement passengers whose destination was the Jivaro settlement of Sucua. Standing on the tarmac, Adriano watched the small aircraft bank to the north and slowly become a small blur as it gained altitude, finally vanishing into a massive cloudbank to the east. He was concerned and miffed that I had failed to arrive in Cuenca. He had taken valuable time away from his studies to arrange for the trip and was outwardly insulted by the pilot's reaction to our cancellation. He knew that if Atessa labeled us unpredictable and risky clientele, we might not be able to utilize their services in the future. He was also concerned about our advance team in Sucua. What would they do?

He continued asking himself these questions as he drove downtown towards the Municipal Court of Justice. He realized that I might be on the next flight and made a mental note to return to the airport and meet the last flight due in from Guayaquil.

Entering the Cuenca airport I immediately spotted Adriano standing near the ticket counter with his arms crossed, a stern look on his face. He pointed to his watch as if I needed a reminder. I knew immediately what was on his mind and what would translate into a long, long day for me. We were two opposites when it came to anger, I preferred to blow like a volcano and put the problem behind me as quickly as possible. Adriano, on the other hand, was a volcano building its dome and would vent slowly, unless really pushed.

He was into phase one of his version of anger management. The drive back to his office was silent, which fueled the guilt slowly creeping over me. Once I apologized and explained what had happened he reminded me again that we missed our flight and that the boys would be at the airport waiting, concerned and confused. The terse scolding was followed by more silence. I refused to take full responsibility and asked him why he didn't take the flight without me. It was checkmate time. He relaxed and suggested we go have a long lunch to plan our next move. Ah, the games people play. But, quietly, in the solitude of my mind, I wondered why he had stayed in Cuenca.

Over lunch we decided to contact Atessa and if necessary beg them for a flight the following day. To show our sincerity we would offer them a bonus explaining the problems and why we were no-shows on the charter. We both felt comfortable that they would be receptive. On the way out of the restaurant, Adriano told me that his family was back from vacation and they were excited about meeting me when we returned from the Oriente. "They think we are in Sucua and are going to be surprised to see us," he said smiling.

Strangely the wrought iron gate was locked. We rang the bell to alert the maids upstairs that someone was at the gate. One of the young maids craned her head over the balcony to see who was calling. When she recognized us she began screaming something in Spanish then quickly ran off to another part of the house. The maid's frantic reactions made us wonder what was going on. We stood staring upwards at the second floor. Rapid Spanish came from the maid and others inside the walled fortress. Adriano repeatedly pressed on the bell and yelled for someone to open the gate. Suddenly all muffled conversation ceased and a strange hush came over the house. We looked at each other wondering what was going to happen next. Then, in rapid succession one head, then two, three and four heads popped over the railing of the balcony. I recognized the two maids and I assumed the elderly couple was Adriano's mother and father. Everyone stared blankly down at us for what seemed an eternity.

Finally Mrs. Vintimilla let out a high-pitched squeal and vanished from sight. The gate latch was released and down the stairs came the older woman mumbling something that sounded like a prayer. She immediately grabbed her son and began crying. The maids lined up behind her started crying. The man, whom I assumed was Adriano's father, wiped away tears and grinned. The Spanish was too rapid for me to understand so I nervously stood by waiting for someone, anyone to translate. Finally, his mother stopped shaking and began to compose herself. She approached, grabbed me by the shoulders, hugged me tight and kissed me on the cheek. It was during this moment that Adriano spoke in a quivering voice that sent shivers up my spine.

"Lee, we are supposed to be missing. Perhaps dead! My family heard on the radio that the Cessna 180 disappeared and the authorities believe it's crashed in the jungle. The airplane did not arrive in Sucua. They will begin searching for it tomorrow."

This day grew into one of the longest days in my life as the four of us sat in the dining room analyzing how fate had intervened in our lives. Mrs. Vintimilla sat next to Adriano, holding his hand, patting it gently. Mr. Vintimilla sat in his place at the head of the table warmly smiling, listening to his son explain the unusual circumstances that led to our missed Atessa flight to Sucua. He began with Jim Bell's stubborn refusal to fly to Cuenca with me. Then he explained my mysterious bout with dysentery and finally my struggle and failure to gain passage on a flight out of Guayaquil.

Manuel Vintimilla Crespo was a big man with rosy cheeks, silver hair and clear eyes that penetrated the soul. His sophisticated appearance commanded respect and exuded wisdom. When he spoke, always in perfect Spanish, everyone stopped talking and listened with respect. He spoke of God's way of allowing us to find our own place in life through other people's tragedies. He said firmly that God's intervention and today's experience should be a learning path for us, and that, perhaps, it was time for self-realization to enter our thoughts and prayers. I was struck with his sincerity and reverence when he described how God had intervened on this day. He said, "The Supreme Being does not waste his time on those who fail to understand that a higher purpose lies beyond. Those who know this are given a second chance in life".

The maids hovered in the hallway, listening to every word of the conversation. Then the phone rang and broke the reverence of the moment. A maid answered the phone and yelled, "Senor Lee, telephono." Surprised that I would be receiving a call, I excused myself from the table. It was Barbara Marsh. Hearing my voice, she muttered, "Thank God."

She had heard on the news that a charter Cessna had crashed on its way to Sucua and knew that I had chartered a plane. She was calling the Vintimilla's to see if they had heard anything about the plane crash. She couldn't believe she was actually speaking to me. "My God, Lee I thought you were dead," she yelled into the phone. "Don't you ever do that to me again!"

The news of the plane crash made the front pages of Guayaquil's daily paper, the Universio, and Cuenca's Mercurio. The somber article discussed Jorque Merchand, the young pilot of the ill-fated Cessna 180 and his four unknown passengers, who were believed to be explorers. The article went on to say that in the past five weeks, turbulent weather in the Andes and the Oriente had been the cause for aviation disasters that had taken the lives of ten people. Five died in an army helicopter crash and now five in the Cessna 180. A photograph of the pilot's grieving widow and two small children accompanied the article. The authorities had already given up hope of finding the wreckage and were assuming that everyone had perished because, as in the past, the dense jungle in the Oriente had swallowed up planes and their crews.

In the early afternoon our men arrived in Cuenca. They had been waiting at the Sucua airport when news of the aircraft's disappearance was relayed to the ground crew. They too, thought that Adriano and I were on board the Cessna that had vanished in the dense sea of green. They canceled everything they had in the works and finagled their way on board a DC-3 cargo plane hauling fresh cut beef from Sucua to Cuenca. They assumed that the Mejia project was finished and they were inbound to show their respect to the Vintimilla family. When they arrived at the Vintimilla residence and caught a glimpse of us their faces turned white. Oriolfo began to tremble and Alfredo looked away. We were ghosts to them. We reassured them we were alive and well and began a debriefing that lasted for hours.

Both were famished and suffering from numerous insect bites, but with knowledge that the project was still on, Alfredo suddenly had animated stories about their six-day expedition. Oriolfo remained quiet. He seemed more reserved than usual, which I attributed to the fact that we were alive.

Alfredo elaborated on their journey up the north bank and their accelerated push, covering five miles, or a little over ten kilometers, in two days before a river blocked their path. He pulled out a map and showed us the defiant river that had stalled their progress. The river was named the Tunanza. They were stalled in no man's land and didn't know what to do. Debating the situation, a Jivaro Indian named Esus, came up the trail and stopped to talk with them. He was the brother-in-law of Antonio Necta, a wealthy Jivaro who owned the house and the land at the confluence of the Tunanza and the Tutanangosa. Our team was on Antonio

Necta's land. Alfredo realized this and asked if we could obtain permission to cross. Esus didn't commit. He only explained that the Necta ranch was named Wakani and Antonio spent most of his time hunting and fishing. He didn't know if Antonio would be interested in working with us. Esus hinted that Antonio might enjoy an adventure; he liked to explore and had a large trunk in his house full of colored stones that he had found in caves upriver.

Alfredo asked if he could see the colored piedras (stones) and Esus adamantly denied the request. He explained that Antonio kept it locked and that he was deep in the jungle and would not be back for a week or two. While the two talked on the bank of the river Oriolfo noticed a mountain without vegetation up river. He could also see several landslides in the same area. He checked his map and estimated that they were about six to ten miles from the Santa Rosa River. They asked Esus if he knew where the Santa Rosa River was? Esus told him it was up the Tutanangosa. Oriolfo asked how far and the reply was "lejando", which meant some distance. Then he asked if Esus had ever been to the Santa Rosa? The Jivaro said that he had not gone to that river, because of his bad leg, but Antonio ventured up there often. The rio (river) was muy peligroso (very dangerous) and there were many "tigres", (jaguars) in the area.

"Only Shuara go there," he said.

Alfredo asked the Jivaro how his relative made his money? Esus refused to answer him. Alfredo realized his questions were too direct. He tried a different tactic, asking what the colors of the stones were that his relative collected from the caves? Esus responded to this question by saying, "shinny".

"But, the color of the shiny stones?" Alfredo pushed.

All colors answered the Jivaro. The Jivaro then turned the questioning back to Alfredo, asking him what they were looking for?

"Gold," Alfredo replied.

"Then why are you asking me about colored stones?"

Alfredo replied that his associates are interested in everything, even shiny piedras. He was angry and convinced that Antonio

Necta had found the Mejia emeralds. In his mind El Dorado was slipping through his fingers.

Alfredo excused himself for the evening, but Oriolfo stayed. We asked Oriolfo for confirmation of Alfredo's story. He was still reserved, but nodded an affirmation. He seemed to want to share something else but was reticent to do so until we began to press him. It was obvious something was bothering him. It was also obvious Oriolfo didn't want to rat on his partner, which made us push even harder. He hesitated but finally explained that the Jivaro did not trust Alfredo because he was Spanish. The Jivaro hate the Spanish. They do not trust the Church or the priests who man the missions in the region because in their view they are Spanish. They felt they have been lied to for the last four hundred years. Oriolfo said the Jivaro's land was being systematically taken away by colonials (outsiders) and the Church when Esus pointed out with disdain that the Salesian mission owned all the land on the south bank of the Tutanangosa and would not allow anyone on their property.

That statement aroused our suspicion and led us into a deep conversation and the possibility that the Church may have found the emeralds and was using this land as a buffer to keep curious explorers out of the area. Adriano refused to accept that theory. He stuck to his guns, staunchly convinced that the Santa Rosa River was on the north bank. The Church property was not blocking exploration of the north bank but, if by chance, he was wrong about the location of the river and if the Church had found the mine, they had workers; a lot of workers. A secret of that magnitude could never be kept quiet. Someone would talk and the government of Ecuador would immediately get involved.

I suggested that they bought and cordoned off the land so they could conduct their own search for the stones. Rumors and stories I'd heard indicated the Church had been conducting their own private expeditions. Several Colombian groups had searched as well and word travels fast in Sucua. I wondered about Crespi. He belonged to the Salesian order and found the emerald encrusted matrix. If there was a conspiracy, was he involved?

Adriano's faith in his Church was strong and for anyone to suggest that the Salesian or Jesuits, his favorite order, would stoop to clandestine entanglements was beyond rational thought. I backed down, but something deep inside was telling me that we were not the only ones searching for an outcropping of high quality gems. Adriano was an idealist, as was I. We were young and just learning that subterfuge was a way of life for many.

Oriolfo listened quietly while we debated the Church issue. I noticed that since Alfredo left the meeting, his demeanor changed. Was there something else he wanted to tell us? I asked him. He looked down and then looked me straight in the eye. Adamantly he stated that he would not go back with Alfredo, unless I was with them. Adriano was shocked by the remark. He perched himself on the edge of the desk trying to appear collected before he pressed for more information. He cleared his throat and straightened his tie, two nervous reactions. Then, prepared for Oriolfo's answer, he asked "Why?"

Oriolfo sat straight in the chair, head up and said that he felt Alfredo was not a good man and could not be trusted because he was Spanish. A major internal problem was developing and Adriano reminded Oriolfo that the Vintimilla's were also Spanish. Then Adriano asked if the basic premise was that all Spaniards are not to be trusted?

Realizing that he had stepped over the line, Oriolfo backtracked saying that the Vintimilla's had always treated he and his wife, one of their maids, with respect. Alfredo had shown no respect for Oriolfo as a person. Oriolfo's eyes welled with tears. The caste system of Ecuador was a sensitive issue. Adriano saw the young man's emotional response and moved to his chair, relaxed, took a deep breath and offered an apology for questioning his loyalty.

"Tell us what happened and why you don't trust Alfredo?" Adriano asked softly.

The angry young man explained that instead of renting two horses for the first day's journey, Alfredo rented only one for him and made Oriolfo walk in front. He was humiliated and his demeanor punctuated the horror story relayed to us. On the first day he was almost bitten by a five foot snake. He'd stopped on the jungle trail to wait for Alfredo to catch up, when he felt something soft under his right foot. He looked down and there was an ugly brown snake squirming, trying to strike. Luckily the snake's head had been pinned to the ground by the Oriolfo's foot, but it was working itself loose. With all his strength he jumped up and backwards as the snake struck barely missing his thigh.

The next five days were even worse. After the first day on the trail, they turned the horse back and continued on foot. Alfredo made him carry both backpacks. When they returned to Sucua, Alfredo gave him some money to go to the small pension and rent a bed for the night. He went to the small hotel and rented two beds, one for Alfredo and one for himself. When Alfredo found out that he would be sleeping in the same room with "an Indian", he became furious and demanded Oriolfo leave the room. The Spaniard was treating him like his personal slave and Oriolflo rebelled and screamed at Alfredo that he did not work for him. He told him that he worked for Senor Lee and Senor Vintimilla. Alfredo laughed at him and said he was nothing but an Indian and not to question his authority.

Adriano and I sat in silence dissecting Oriolfo's story, which hadn't finished. He told us that they had washed gold in the Tutanangosa and had brought back several ounces, but Alfredo took charge of the precious metal and made Oriolfo swear to him that he would not tell us about the gold. The Canari said that he was forced to swear because he was afraid of Alfredo, especially when he started drinking, which was every evening. He said he was happy that they did not get to the Santa Rosa River and find the emeralds. He felt that this 'mad man', as he referred to Alfredo, would have killed him. I patted him on the shoulder and thanked him for his honesty and courage. I knew that it took guts for him to express his concerns.

Because of Necta's wealth, Alfredo was convinced the Jivaro had found Mejia's emeralds. Oriolfo said that Esus confirmed that Antonio was a wealthy man, wore nice clothes, and had two wives, a home in Sucua near the airport and a ranch named Wakani, on the Rio Tunanza. He owned a bull, many cows, and he drank good liquor. He was an honorable man who carried the title Don in front of his first name and was revered by the Jivaro in the region as a trustworthy friend.

"If he likes and trusts you, then he will allow you to pass through his land and maybe even work with you. But you must earn the trust," Esus told Oriolfo. Esus and Oriolfo developed a strong friendship those few days in the jungle and Oriolfo had gained much insight to the mystery man, Necta.

Don Antonio's wealth had become Alfredo's Achilles heel. In the mind of Alfredo there were only two sources of wealth available to the Indians of the Oriente, either gold or emeralds. Since Wakani was located near the Rio Santa Rosa, it was obvious, at least to Alfredo that Don Antonio's wealth came from emeralds. After Oriolfo's explanations we knew for certain that Alfredo was a major problem. His greed, the north bank and its many landslides and unpredictable rivers awaited us.

Another briefing had been arranged for the next afternoon, but Alfredo didn't show at the designated time. An hour late, Alfredo entered the room with his head down. He smelled of whisky. Adriano was furious and asked him point blank, what his problem was? He said that he had some personal problems. When asked about his treatment of Oriolfo, he dismissed it as exaggeration from the young Indian, saying that Oriolfo should have never been selected to go with him into the Oriente, because he was young and inexperienced for such a difficult trip. He refused to discuss the gold, saying that it was an insignificant amount and of no importance which infuriated Adriano even more.

His arrogance pissed me off, so I told Adriano to translate the following. "Since our team was unraveling because of internal discord, I would no longer fund the project and I was pulling out unless things changed dramatically." That said, I stormed out of the meeting and went to the dining room and poured myself a cup of coffee and waited for a reaction.

I could hear Adriano's voice rise with anger as he echoed my sentiments. Alfredo excused himself and stumbled into the evening an abandoned and lonely man. Adriano, still furious over Alfredo's actions joined me; he was so mad he trembled. This was the first time I had seen him this way.

Oriolfo stood in the doorway, took a deep breath and confessed that he was afraid Alfredo would try to kill us for the emeralds. I didn't want to confirm his suspicions, so I handed him a cup of strong steaming coffee and offered a half reassuring smile. It was that crazy look in Alfredo's eyes and his total lack of integrity that alarmed me. My thoughts went to Barbara's 'Bogey' analogy.

It was best for me to keep my concerns private until a favorable solution could be worked out with Alfredo. I assured my friends that Alfredo would be back, hat in his hand, once he sobered up and realized that he needed us more than we needed him. Adriano asked if I was serious about pulling out of the project. I assured him it was only a chess move to counteract Alfredo's stupidity. He relaxed and headed for the study to hit the books and prepare for the most important three weeks of his life.

I was outwardly brimming with confidence that my dramatic exit had been effective. I hoped Alfredo would remember the evening's events. On the inside I was churning with doubt as to whether I wanted to go into the womb of the Macas Forest with an idiot that was slowly losing his grip of reality? My thoughts returned to Barbara and her brilliant dissertation on greed and my ability to defuse an explosive situation before it became life threatening.

Was I over reacting? Or did we have a real "Bogey" on our hands? I knew it would be easy to remove him from the project, but if we did, the project's secrecy would be compromised and he would be undermining us at every turn. I felt the only plausible option was to keep him on board, on a short leash, working in the direction of teamwork, even if it took a swift kick in the ass every now and then. The bigger problem might be trying to keep him off the sauce. I had been around drunks. My mother had married one and I had no tolerance for their uselessness.

It was a little after four a.m., when my bed shook with such force it almost tossed me onto the floor. "My God it's an earthquake," I thought.

I steadied myself placing my hands on the wall next to my bed. The violent shaking lasted a few seconds. I expected the wall of the old house to cave in at any moment. Then, as suddenly as it started, it stopped. My eyes finally focused and in the bright moonlight shining into the room, the wall, floor and ceiling did not appear to be moving. I stumbled to the light switch and flipped on the lights. "It had to be an earthquake." I thought.

I opened my door to see if anyone else was preparing for a sleepless night in the street. The house was dark. There was no panic, and nobody was stirring. Immediately, I went to the wall next to my bed and examined it carefully, looking for cracks in the plaster. There were none. This had to be one hell of a bad dream. I crawled back between the covers and forced myself to let it go and soon went back to sleep.

The next day Alfredo called as we were sitting down to dinner. He asked if he could come by for a few minutes to offer an apology for his outlandish behavior. We set a time and, as I had predicted, he arrived, hat in hand. We verbally accepted his apology; however I knew I could never trust him again. Before leaving he gave Adriano a hug, shook my hand but never made eye contact with me, and left the last will and testament stating that it would be safer with us. When the door closed behind Alfredo, Adriano grinned from ear to ear. He suggested I get a degree in psychology while he finished his law courses. I laughed then asked if he felt an earthquake last night? He said that he hadn't felt anything. I didn't respond, but I couldn't help thinking that maybe I needed the psychologist.

The following nine days became a lesson in patience as we waited for the weather to clear. The down time or dead time as we called it, was excruciating and frayed the nerves as we waited and waited...When it was raining in Sucua, the weather was clear in Cuenca and vice versa. Unfortunately, there were only two practical ways into the wild, west settlement of Sucua—air and air. The Ecuadorian Military Air Transport, TAME, which flew vintage DC-3's and the other by a private charter service named Atessa, which, was owned and operated by Captain Drexel. These were our only options and both carried risks.

Drexel was a folk hero in Ecuador because of his uncanny ability to survive the hostile skies over the Oriente. The cagey captain was rumored to be an ex-Stuka pilot who flew for the German Luftwaffe during World War II. Unfortunately, the disappearance of his Cessna 180 meant that one-third of his air fleet was gone. Over the phone he assured us that his super Skymaster was now on permanent location in Cuenca and would be available for us. But, he cautioned that we had to be prepared to leave within an hour's notice due to of the unpredictable weather between the two locations.

On the fourth day, the telephone rang. It was our time to fly. While Adriano located Alfredo, I double checked my backpack and ran a quick checklist of our supplies: canteens of fresh water, snake serum kits, and food. With Merchand's fate fresh in my mind I added an extra bandoleer of 20 rounds for the 357-magnum pistol. Although we were only going in for a quick three-hour turnaround, I was preparing for the worst. If our flight should go down and we lived, survival would depend on us walking out. We couldn't expect to be rescued. The Army still had not found the first missing Cessna.

We jumped into the Land Rover and headed to the Cuenca airport where the captain and his blue and white Cessna Skymaster sat waiting on the edge of the tarmac next to a small hangar. As we unloaded our gear the Captain introduced himself, scrutinizing our backpacks and us. He was a tall, rugged man in his late fifties, with light brown hair and a clean cut Aryan face that substantiated his German heritage. A man of few words, his physical appearance would easily have qualified him for poster-recruitment for the German Luftwaffe. His demeanor was strict Prussian in manner and style. Adriano climbed in front and Alfredo and I crawled in back. Once settled, Drexel commanded us to strap ourselves in for the thirty-minute white-knuckle flight. The Captain's background became a mute point as the Cessna screamed down the runway under full throttle, grabbing frantically at the thin air, trying to get airborne.

Once up, The Cessna was a flying machine of beauty and power. We circled to gain elevation above the 12,000 foot mountain peaks that stood as guardians to the Oriente. The altimeter showed 11,000 feet and we were skimming mountain ridges at 140 miles per hour. Below I could see the murky Rio Paute snake its way east between towering ridges covered with thick jungle. There was no radio beacon for navigation and the brave pilots flew through jungle covered mountain passes with only a clock on the wheel to determine when to change headings. His navigation was broken down into minutes and seconds; it was a lesson in sheer terror on a cloudy day.

Today luck was with us. Visibility was perfect and we crossed the cordilleras and descended into a smooth flight, above a sea of endless green rain forest and mountaintops. The Captain seemed curious about me. He turned around, eyed me up and down, then asked, "Are you Jewish?"

I shook my head. "Nope, I'm English-Irish."

A crooked smile crossed his thin lips and he turned his attention back to the aircraft. Looking out of the window at the rough terrain, I wondered if he would have thrown me out of the plane if I had answered in the affirmative.

Before long, the cabin radio came to life with the constant chatter of Spanish from Army Radio Sucua asking our identification. Permission granted, we landed on a small, grassy, washboard runway in the Upano Valley, the home of the settlement of Sucua. Drexel unloaded sewing machines, radios and small boxes while the three of us made our way down the dirt road to the Jivaro settlement.

It was a hot day. Adriano and I wore jackets to conceal shoulder-holstered weapons under the heavy backpacks crammed full of survival supplies and gifts for the Jivaros. We were miserable during the mile and a half walk to the outskirts of Sucua but I saw hundreds of beautiful butterflies darting in and out of the foliage, which made the discomfort more tolerable. We finally reached a thicket of banana, grapefruit and orange trees mixed in with ugly Wyacan trees.

A fierce white dog dared us to try and enter his domain, a thatched house sitting on stilts. He was about to attack when a small, muscular Jivaro man stepped from beneath the house and positioned himself between the dog and us. Shouting a harsh command in a strange dialect, the guard dog snarled and retreated. Alfredo introduced us to Esus, the Jivaro he and Oriolfo had met on the trail near the Tunanza River and the son-in-law of Antonio Necta, the man we had come to see. Esus had straight jet hair, high cheekbones and slanted, wild eyes. He was friendly and invited us into the house. We followed him and I noticed he limped as he walked.

We were disappointed to find out that Antonio Necta would not be joining us. He was in the jungle and was not due to return for several more days, but Esus offered to give us any information that he could. Setting up my tape recorder, I noticed Antonio's wife studying my every move. She whispered to Esus. Alfredo said that they had never seen a tape recorder and were wondering what the machine did. Once it was explained to them they seemed at ease and then asked if they could record their voice in the box before we left. I assured them that I would be honored if they would say something to the box. They had no idea that their discussion was being recorded as they chatted between themselves. I asked if they'd like to talk into the machine when we finished and they giggled like school children.

Alfredo had warned us earlier that the Jivaro were very jealous and possessive of their women and not to have direct eye contact as this might be considered flirtatious and could create problems. When Antonio's wife looked in my direction or spoke to me, I found myself avoiding eye contact by looking at the ground or away, which was very uncomfortable for both of us. She was very unattractive and old for her true age. Her daughters sat in the corner whispering to each other. It was a struggle not to look at them because of their sensuous mixture of wide-eyed innocence that blended exotically with primitive beauty.

For the next hour we sat cross-legged on the wood plank floor adjacent to the open cooking hearth and discussed many things. Strange sounds coming from a strand of trees next to the house kept distracting us. I felt like we were being watched.

Interestingly, the Rio Ojal kept entering the conversation. Esus spoke of a tree in that area that produced a "juice" that was used by Antonio as a cure for an American doctor who had visited them in the past and was suffering from a strange disease. He could not remember the name of the doctor or the disease. "You will have to ask Antonio about this," he said.

He also mentioned that near the river Ojal there is a cave that has no end to it and the front of the cave shines when the sun is directly in front. There are "muchas piedras colores," (many colored stones) he said. "The Shuara do not like to go there, because there were no dry leaves at the Ojal to build fires at night and the tigres are very aggressive."

When the Jivaro spoke of tigres, they were referring to the jaguars. When they mentioned tigrillos, translated as little tigers, they were talking about ocelots. The pantera I knew was the puma and oso negro was the black bear. The most feared were the jaguars because of their size, but the meanest were the ocelots that used their razor sharp claws to hold and maim their prey. On my two previous expeditions, I had never given a thought to four legged predators, only the slithering kind and the two-legged variety. Esus explained that the tigres at the Ojal were "muy bravo", very brave. This was a problem compounded by the fact game was scarce in the area. Anything that moved was a possible dinner. About that time the strange sounds coming from the thicket next to the house became louder. I reacted by reaching inside my jacket for the reassurance of my 357. I was wondering if I was about to have my first encounter with a tigre, when once again, Esus made reference to the large cave with shiny colored stones.

"Cascabel?" Adriano whispered to me.

Aside from the mystery cave Esus didn't offer much insight. Before leaving the Jivaro compound, I pulled the Polaroid camera from my backpack and asked if I could take pictures. The Jivaro weren't certain what I wanted, but agreed anyway and I took a photograph that instantly developed before their eyes. It was a huge thrill for the startled onlookers when their likeness slowly came into focus on the photo paper. Their eyes widened and Antonio's wife let out a high pitch squeal when she recognized her own image. When I gave Esus the photo of himself standing in front of his house, he was overwhelmed with joy and bewilderment from this modern technology that allowed his image to spring to life in instant color.

The Jivaro were uninterested in the transistor radios we offered as gifts; instead they wanted more photographs to be taken. I took three more before asking them to say something on my reel-to-reel tape recorder. Antonio's wife approached the recorder cautiously, but after being assured by Esus that it was harmless, she spoke in their native dialect without any apprehension. Her phrase, "the sky is blue" was a jangle of mixed sounds of the unusual dialect, which was impossible for me to repeat. I gave it a bold effort but was scolded by her for my repeated failure to create similar sounding words. I was laughing at my raw and unacceptable efforts when the noise from the thicket began again. I slid my hand under my jacket and gripped my pistol. If a jaguar made an appearance I was going to be prepared. Esus watched me. I could feel his questioning eyes studying my every move. I turned staring right into his smiling face. I couldn't understand why he was amused by my caution.

"Parrots, Senor Lee, only parrots." Esus grinned.

I felt like the proverbial city slicker. We left three packs of Kent cigarettes, four Polaroid photos, two transistor radios and two airplane tickets to Cuenca with the family before leaving. In return, they gave us a sack of oranges and an open line of communication, which we considered a fair exchange.

As we trudged down the dirt path leading to the road and the airport, Adriano kept repeating, "I knew it was the Ojal." Then he quoted the last will and testament, *"It takes crossing the first river upstream, the entrance is near the cascabel."* He believed that the cascabel was the large cave that Esus spoke of that had no end. "I think we are very close, Lee," Adriano said with confidence.

At this point in time he had no arguments from either Alfredo or me. On the army map the first river upstream from the Santa Rosa was the Ojal and the cave or tunnel was near the Ojal. We were chattering back and forth like the parrots in the thicket next to Necta's house over the exciting possibilities that had surfaced during the past hour.

Alfredo, who was beginning to understand English, after weeks of hearing Adriano translate for me, disagreed. "I believe the Jivaro, Necta, found the emeralds. Otherwise, how did he obtain his wealth?"

Adriano countered that Necta may have obtained his wealth through washing gold or even inheritance. About that time he pointed to the side of the trail. Alfredo and I jumped! We thought he was pointing at a snake on the trail rather than another beautiful butterfly sunning itself on a broad green leaf. We were hypersensitive because Esus had told us that there were an abundance of snakes here in Sucua because of the hot and humid weather. But at the Ojal there were tigres but no snakes.

Captain Drexel stood by the Cessna talking with two soldiers from the small military garrison in Sucua when we arrived in our coats hauling a bag of oranges. We must have looked suspicious to the soldiers. What fools would wear jackets in 100-degree temperature. This was a mute point because the captain was doing his job. He was planting our cover story with the military. Adriano had informed Drexel before we left Cuenca that our mission to Sucua was to look at potential farming land that might be purchased or leased and we knew he was perpetuating the rumor when the soldiers smiled and nodded. El Dorado syndrome was alive and well in Sucua on this day.

Our bodies throbbed with vibrations and deafening noise during full throttle takeoff from the dirt runway as the Skymaster climbed into the hot sultry air. The Cessna circled to gain the much-needed elevation and we were able to get a good look at the Tutanangosa and the dense jungle through which it twisted and turned. It was an awesome feeling to know that within this peaceful, almost surreal setting a fortune in emeralds, guarded by the forces of time and nature, lay below. Straining my eyes, I searched for a sign of the Rio Ojal headwaters and a cave, but my vision blurred in the sea of green.

At the Vintimilla residence we poured over notes and listened to the tape recording searching for hidden meanings and clues from our one-hour meeting with the Jivaros. Antonio Necta, according to his son-in-law had broken off pieces of the colored rocks in the endless cave and was paid two hundred sucres for the shards. That amounted to only about ten dollars, which was nothing compared to our standards, but was a lot of money in the Oriente. Our main question was what were these stones? It was certain the Indians in the Oriente had no idea what constituted a precious stone. They collected colored stones because of their shiny beauty not because of possible value. Esus had mentioned that Antonio had found five different types of colored stones: brown, yellow, blue, pink or red and green.

Necta's lock box held our curiosity. We had listened to Esus discussing the two Jivaro men that had broken into Antonio's house trying to steal the cache of colored stones, but they were discovered and had fled for their lives. We understood why the white dog was stationed near the house and always in the attack mode. He was the appointed guardian of the lock box and anything else of value in the house.

It had been a long, tiring and somewhat emotional day and I couldn't wait to take a shower. Insect bites suddenly came to life in the warm water and were beginning to itch unmercifully and transform into large red welts on my legs and arms. I realized why Adriano had been scratching during our flight home. After treating my bites with fingernail polish remover in case they were chiggers and not mosquitos, I decided to cut a tape to the guys in Phoenix and bring them up to speed on my first trip to Sucua.

I ended the tape by requesting more firepower. We would be going into an area that was teeming with jaguars, pumas and ocelots and I pleaded for another rifle, automatic shotgun with 00 load cartridges and more Polaroid film to sooth the savages.

"Forget about the transistor radios. They're not interested in them. Batteries are hard to come by and very expensive over there. And please guys," I begged "If you can, send me some sports news, especially about the Detroit Tigers. Did Lolich, Kaline and Willie Horton make it into the World Series? Did they win the Pennant? Also, guys" I crowed, "Cuenca now has its first television station and the top three television shows are Combat, Superman and Peyton Place."

I could see the bright light emitted from the converted sitting room, now the TV room, where the Vintimillas and the maids sat in total ecstasy watching a black and white mini-screen bringing them the magic of Hollywood entertainment to be enjoyed in the privacy of their home. Cuenca was slowly, very slowly, moving into the modern age.

I scribbled in my diary: *It had been a most interesting and surreal day! Beginning with a nervous flight, on a charter service still suffering from the loss of an aircraft and pilot that had our names written all over it. Flew into a Jivaro community, where we bartered American cigarettes and Polaroid film for information and returned with oranges in time to watch Superman speak in a husky, macho Spanish voice about the perils of evil. God, it was great to be alive in the fall of 1968, to witness technology moving by leaps and bounds. Now if they could only get the telephones to work in this country and discover how to make a hamburger and fries that tasted like the real deal, then I'd have it made. What more could you ask for? A football game would be great and even better the World Series. But that was wishful thinking. This was the land where communication stood locked somewhere between Inca runners and the pony express.*

The first week in November our dead time was occupied by waiting for the Jivaro to arrive in Cuenca. I don't wait well. The days were rainy and long and the nights cold and longer, I amused myself by studying Spanish and re-reading the last will and testament of Raphael Mejia. The highlight of that long week was when Adriano invited me to join he and Sonia for a movie. I was looking forward to meeting the girl that had stolen his heart and mind and I couldn't wait for Wednesday evening to roll around. Her name was Sonia Ugalde Puyol, an attractive, elegant, tall, dark haired girl that had a commanding presence about her, whether she was sitting next to you in a theater or holding court in the Vintimilla living room. She was a very special person and I understood why my friend would soon be headed down the altar with her.

We went to see the film everyone in Cuenca was talking about, Sergio Leone's, "The Good, the Bad and the Ugly" starring Clint Eastwood. The theme song of this film, which included a whistling medley, was reenacted for weeks on end, especially late at night, when the movie-goers walked down the cobblestone streets, headed for home. The sound of their whistled rendition echoed through the stillness of the night cascading against stone buildings and cobblestone streets. It sent shivers up my spine. Many a night, almost asleep, the whistling would jar me back to semi-reality with visions of Eastwood dancing in my head. Hollywood won over the rebellious students when politics failed and this was apparent when Clint came to town, even if it was just in celluloid. The students fought each other over our involvement in Viet Nam and other pointless politics, but "The Good, the Bad and the Ugly", became their theme song and for two hours when they forgot about politics and allowed an American actor to become their hero.

The evening before the Jivaro were due to arrive, I went to bed early and spent an hour scanning the broadcast bands of Armed Forces Radio and Television, (AFRTS) and the Voice of America, trying to find the results of the U.S. elections and the World Series. A weather inversion had blocked out the signal strength from these stations so I was left with two choices, the BBC, which was paying tribute to Mozart, or Havana Anna who was coming in strong from Cuba.

I settled on Anna the firebrand with the sensual voice, which really didn't match her topic of the evening. "Decadent American women who wore nylons and painted their faces and made them look like clowns in a traveling circus." She continued to praise the Vietnamese women for sacrificing their virtues in their struggle against the American war machine. She began to wear on me when she began reading the week's American casualty list, which included names, ranks, ages and serial numbers of those killed by "the patriotic artillery fire of the Vietnamese forces." It was time to turn off the light and get some much-needed rest.

I tossed and turned in my small bed unable to block out that damn casualty list she had read. Most of the soldiers she mentioned were young guys in their teens and early twenties who had never had a chance to experience life, love and the pursuit of happiness. My purely philosophical thoughts centered on the age-old question: Why did some live and others die?

Perhaps reincarnation was the answer? This subject was new to me but I had read a few books, even though it seemed like some ancient Eastern mythical philosophy. The concept opposed my staunch Christian up bringing, but somehow, in some strange way, it made sense. I found myself struggling with the differences, even after Adriano told me that the Jivaro also believe in life after life. If we return in each separate life for the purpose of learning, then have many of us died in warfare at some point in time? I read somewhere that General Patton believed he was a Roman soldier in a past life. I guessed that meant he was drawn to war, but what about those of us who saw it as senseless? Does that mean that those of us that learned the lessons of war have no further need to participate in this form of bloodshed? Perhaps this was why so many young Americans fled the country rather than be drafted for Nam? If reincarnation was real, this could mean that many of the draft dodgers weren't cowards. Instead, maybe, they were old souls that previously paid their dues and learned their karmic lessons and wanted to move forward? It was obvious, to me at least, in the most secret spaces of my heart and mind that I had a lot of learning to do. I was still wrestling with religion in general, reincarnation in abstract, and mountain spirits in afterthought.

I flopped around in the bed before my mind focused on another surreal question. How does one fight in a jungle? The jungle itself is a formidable foe, with its own forces that defend their turf against any attempt of intrusion. Its forces come in all shapes and sizes, from microscopic bacteria, fungus and amoebas, to soldier ants, poisonous spiders and snakes. Then there are the predators, the tigres who can weigh up to three hundred and fifty pounds and the mean little ocelots that can tear a man to shreds.

The major insurmountable foe within a jungle was the unpredictable weather. Rain is constant and everything gets wet and mildewed including weapons, food and clothing. But no one knows how much it will rain or if the rain will manifest a flood. Combine all of these forces of danger and then add a Vietnemese guerrilla force of combat seasoned troops with AK-47's, that you can't see, hear or feel and you have a no win scenario. My heart went out to our fighting forces in the green hell of Nam, who like myself, perhaps were still wrestling with the wheel of karma, even if they didn't recognize it.

Still tossing and turning at 2:00 a.m., my thoughts finally slowed and I drifted into sleep. My sleep was soon interrupted by the creaking sound of the loose lower step of the staircase. I heard soft footsteps coming up the stairs, which were located next to my room. My first thought was that Adriano was returning late and was trying not to wake anyone. In my drowsiness, I checked my watch and it was 4:00 a.m. It was unusual for him to be up this late, especially with the heavy day we had planned. I heard footsteps on the tile floor leading to my bedroom door, then up to the door. They stopped. The mystery person was standing at my door, silently blending in with the darkness and stillness of the old house.

I was wide-awake! Trying not to make any noise I reached for my flashlight and pistol that were in my backpack under my bed. There was an intruder in the house who was checking out my room. I located the flashlight but was still fumbling for the weapon when the wooden door buckled from the force of a blow. The sound had to have sent shock waves through the stillness of the house, but no lights came on. Everyone was still sleeping when I located the revolver. I approached the barred door with flashlight in one hand and gun in the other, when the door received another resounding blow from someone...or something on the other side. I was really pissed, high on adrenaline and ready to do battle when the light from my flashlight went out. Crouching in front of the double set of wooden doors, I quickly lifted the wooden plank that acted as a dead bolt and swung the doors open. Flipping on my room's light switch I pointed the cocked pistol at the intruder.

There was no intruder. The terrace area was empty. There was no sound of anyone running or trying to escape. Nothing! Even more bizarre no one was stirring in the house.

Retreating into the bedroom, I found fresh batteries for the flashlight and reloaded it on my way to the stairs that led to the roof. The intruder must have entered the house from the roof, I thought. The door and gate downstairs were locked, alarmed and electrified. I opened the door to the roof terrace shining the flashlight into shadowed areas searching the dark perimeter of the building. The bright moon aided my efforts but there was no one on the roof. After a ten-minute search, I returned to my room still puzzled and alarmed over the strange events that had taken place. Had I been dreaming? Was I hallucinating? Did the event actually occur? How did the intruder escape to so quickly? And why didn't someone else hear the crashing sound when he or it rammed the door of my room? A myriad of questions flooded my mind until I finally drifted towards sleep as the early morning sun began to filter through the terrace skylight.

The next morning over breakfast, Adriano's mother and father listened quietly as Adriano discussed the Jivaro trip to Cuenca. Since returning from their summer vacation they had been enthralled with the Mejia project and had become our ultimate supporters and confidants. Virginia Ordonez de Vintimilla was a beautiful woman in her mid-fifties that always displayed a radiant smile and willingness to help others. The entire family had adopted me and they always made sure that I was well fed and comfortable in my surroundings. Mrs. Vintimilla was the first to notice how quiet and distant I was during breakfast and unable to speak English, she asked Adriano to find out if I was feeling ill.

"My mother says that you don't look rested and your face has very little color, she asks if you are feeling well?"

I replied that I was tired because I didn't get much sleep last night. Adriano translated to his parents.

I was asked if something was bothering me. "Well..." I paused while searching for some kind of rational explanation. "Yes, there is, but I don't know how to explain what happened last night without you and your family thinking that I am going nuts, but I will try."

For the next few minutes, I gave the family a detailed description of the unusual events that had occurred in the wee hours of the morning while Adriano translated word for word to his parents and the maids standing in the doorway. I had expected to see everyone wide-eyed and staring at me strangely, instead, Adriano's translation was punctuated with his patented, stifled laugh while the family disarmed me with their perpetual all knowing smiles.

Adriano laughed harder. "Lee, you are not going crazy, we've all experienced the strange things that occur in this house. The house is over one hundred years old and like many of the others in this city has a history of mysterious hauntings. We believe there are poltergeists in the house."

"You mean ghosts?" I asked nervously. "I thought ghosts could walk through walls and doors, not try to break them down."

"We call them poltergeists, and yes, we've seen wispy apparitions in hallways and bedrooms. Sometimes we hear coins being tossed and rolled down the stairs. We hear a turkey gobbling, too. That comes from the downstairs storage room and sometimes a woman screams. When we go and investigate the sounds cease and we find nothing. The lights in certain rooms sometimes go on and off mysteriously, too."

By now the family had joined in with their own stories and Adriano was hard pressed to keep up with the many experiences with the mysterious entities. "In the past, we had problems with hired help who were always quitting at inopportune times. So my mother made sure that when we hired our present staff they were told of the poltergeist, however, they were also made aware that no one has ever been harmed or endangered by our noisy but seemingly friendly guest. When we invited you to stay at our house, I thought of telling you about this, but was hesitant because I, too, did not want you thinking I was nuts. If I told you and then nothing happened, you would question my sanity. Do you believe in spirits? Or does this type of thing frighten you, my mother asks?"

"No, I wasn't frightened. Last night I was startled and angry that an intruder was in the house...a physical intruder, I might add."

Then I explained my early childhood growing up with my grandmother in a religion that encouraged spirit recognition through visitations by angels and her ability to speak in unknown tongues. I explained how this belief conditioned me to understand the pet coyote that the Apache Kid said was my spirit dog, and later how a mysterious dog joined my expedition to Infiernilios and eventually saved our lives. Was he my spirit dog from the past? I didn't know, but I would like to think so. The family was pleased I had not packed up and moved back to the Hotel Cuenca. They began talking among themselves and the maids about their personal experiences with Casper the friendly and extremely loud ghost. It appeared my dead time was being shared with the spirit world.

CHAPTER 8

THE MYSTERIOUS ANTONIO NECTA

Antonio's wife, Antonio, and Lee in Sucua

Anticipation charged the atmosphere among the domestic workers who chattered like parrots about the pending arrival of the Jivaro. None of them had ever seen a Jivaro Indian before and only knew of their reputation as the headhunters of the Western Amazon. When the gate bell signaled the arrival there was a stampede of workers who rushed to the staircase for an unobstructed view. The Jivaro were greeted at the foot of the stairs by Adriano, who according to protocol, introduced his father and mother, then me and Alfredo. To the dismay of the house staff the visitors were dressed in their finest western styled clothes and shoes and there were no spears, blowguns or shrunken heads in sight.

Antonio Necta was a small man, maybe five-feet five inches tall, with dark piercing eyes that were well proportioned on his round face. Esus had wild, angry eyes and a face that had deep lines etched on it from the hard life of the Oriente. Antonio's face looked smooth and soft for his fifty some years. His eyes sparkled with wisdom and knowledge. He greeted us in a high-pitched voice; when we shook hands his grip was firm and strong and he seemed to hold eye contact longer than normal. I felt he was searching my eyes, my soul, for something. Whatever he was looking for, it was done in such a comforting way that I quickly accepted it without feeling a need to dominate or control the situation.

We had expected only Antonio and Esus to make the journey, but Antonio brought his sixteen-year-old daughter, Maria, who had never been outside of Sucua. It was a gesture that delighted Mrs. Vintimilla, who immediately disappeared into her bedroom and returned to offer the beautiful, wide-eyed girl, a gold locket as a symbol of the family's friendship and hospitality. Maria looked to her father before accepting the gift. When he nodded, she took the locket and cupped it in the palm of her hand, ever so gently like a child protecting a delicate treasure.

Mrs. Vintimilla asked Maria, if she had ever seen television. Maria, who had been schooled by the Salesian mission, replied that she had never seen a movie or television. As she was escorted to the TV room, I glanced at Antonio who seemed touched by the generosity shown to him and his family. I could understand his emotion.

Indians were not welcomed in the homes of the Cuencanos unless they were domestic help. We had tried the whole morning to find a hotel in Cuenca that would admit the Jivaro. The hotels Cuenca and Crespo had refused them. We finally located an "Indian" hotel that would accept them providing we paid in advance and took full responsibility for their behavior and safety. The caste system in South America mirrored that of the deep south in the United States.

Antonio had brought a small black suitcase with him that contained his personal belongings and those shiny colored stones that had piqued our curiosity for weeks. The brown stone, which he offered as a gift for me was an agate and the white, pink, blue and green stones were only colorful quartz. When Alfredo saw me comparing the green quartz with the many quartz crystals in my field manual book on minerals and rocks, his face reflected joy and relief. His anger and greed, which had been magnified by his supposition that Necta had found the Mejia treasure, was now being unmasked. He carefully examined the pictures in the manual. In Alfredo's mind the treasure was still there for the taking. Alfredo's demons may have retreated but Adriano and I were disappointed that the black bag did not contain any green fire.

I was curious so I asked Antonio if he had ever found an emerald? He replied, "No." Adriano asked if he knew what an emerald looked like? He said, "No."

Alfredo did not want to be seen in *his* city socializing with savages from the Oriente. He did not volunteer to take the Necta family to their first movie and left that assignment to Oriolfo whom was honored to undertake the task. While the Jivaro marveled at Hollywood magic that evening, Adriano and I wrestled with a plan of action. It was apparent that we had a new dilemma on our hands; the Jivaro had no visual concept of the precious stones we were looking for and were ignorant of their value. Antonio had collected the colored stones for their beauty and not for their possible value. We had assumed the obvious and now we were forced to rethink our original plan. I felt we must come clean and place our trust in the hands of Don Antonio Necta and tell him about Raphael Mejia and his emerald discovery. After a brief discussion it was agreed that we would show him the last will and testament.

I knew it would not be an easy task to educate these people on the characteristics of a real emerald. Tourmaline, green quartz and a variety of other semi-precious stones often fooled gem hunters. We quickly ruled out the educational process as time consuming and too confusing. We would only ask that they collect any beautiful green stones they should come in contact with and hope for the best. In the meantime we would have Antonio focus on Mejia's clues.

The next morning everyone arrived on schedule including Alfredo who was sober and anxious for the meeting to begin. It was amazing what greed could do to a man. It either kept him high as a kite or in a pit of despair. Maria was wide-eyed from her Clint Eastwood experience but her father and brother-in-law appeared subdued and uncomfortable with city life. Adriano sensed their uneasiness and immediately began to explain the Mejia story in slow Spanish.

Before he could finish, Antonio, who was by now more relaxed and comfortable in his surroundings, interrupted. "I know this story. I heard this story from a very old Jivaro man who was eight years old when the first Colombians came to the Macas forest. They were the first white men he had seen and he ran with fear to the home of Singusha, a shaman to protect him."

I was stunned and Adriano was speechless by Antonio's reference to the shaman Singusha. He had just confirmed a valuable piece of the last will and testament as authentic, otherwise how would he have known about Singusha and his involvement with Mejia?

Antonio Necta listened intently as Adriano recited the Mejia riddle.

"Like making a relation, the entrance is by the cascabel and it takes crossing the first river upstream."

Antonio responded in a high-pitched voice saying, "Senor Lee, Senor Vintimilla, I know the cascabel. It is a tree that grows in the jungle. It is named cascabel because it produces a pod of dry seeds and when the wind blows against the pod it rattles like the snake with the same name."

Adriano and I just looked at each other in disbelief as Antonio continued. "I also know a very old man who will know where the Tambo Santa Rosa was located. If I can find him, I will have him show me the location. The last time I saw him he was very ill and unable to walk very far. But, I will try to locate him and if he cannot go with me, I will ask him to draw a map for me."

Adriano and I tried in vain to control our excitement. We opened the book to the Mejia map. Then Antonio unfurled his well-worn map of the area and another surprise unfolded before us. The two maps showed the Rio Santa Rosa in the same location, directly beneath the Rio Ojal.

"The Rio Santa Rosa has to be the river," Adriano whispered to me. I nodded in agreement and Alfredo winced at the suggestion.

"Antonio, do you know of the rock that is open in the middle?" Adriano asked. Antonio looked at Esus and both paused for a moment before shrugging their shoulders and shaking their heads.

"Will you work with us and lead an expedition to search for this treasure?" I asked.

Before Antonio could answer, there was a knock on the door of the study. Adriano excused himself and went outside to speak with his mother who handed him an urgent telegram. While they were conversing, I watched Antonio trace the Mejia map with his index finger while carefully comparing his old weather beaten map. His eyes sparked with curiosity and a sly all-knowing smile sneaked across his smooth, ageless face. He reminded me of a youngster who had just solved his first jigsaw puzzle and couldn't wait to begin the next one.

"Was the mysterious Don Antonio Necta too good to be true?" I wondered silently. Something about this man was mesmerizing. He had two faces: one he was showing us, and the other, the real face, that he kept hidden from view. He reminded me of the Apache boys I grew up with but Antonio was more at peace within. Whatever he was hiding from us colonials would be given, in time, after he decided we were trustworthy.

Adriano returned to the study, handed me a telegram marked "urgent" and continued his discussion with Antonio. I tore open the telegram as Antonio stated that he would be happy to work with us and, yes, he would lead an expedition up the Tutanangosa and to the headwaters at the Ojal. While Adriano and Antonio were working out the details of the pending expedition, that would hopefully take place within weeks, I was reading the following: "I will arrive next week with new funds and new blood. Best, Jim."

Antonio suggested that it might be best if he took an advance party of trusted friends in to look for the clues and pave the way for a later expedition to be undertaken in December. That way they could tell everyone that I was interested in hunting tigre in the area. This story would throw off any suspicion of treasure hunting. It made sense. December began the dry season. Don Antonio Necta was mysterious and also very resourceful and clever. I liked how he thought.

We shifted the conversation back to the Rio Ojal, which translated to "button hole"; it was not Antonio's favorite place. We asked him to describe the weather, terrain and the caves in the area. Antonio had a soft voice that grew high pitched when he became excited and the Rio Ojal brought out the high pitch squeal almost immediately.

"Many brave tigres there! Once you leave the Rio Tunanza you are in no man's land. There is no trail and many snakes. A black bear lives there and he will attack a man. Black panthers are muy peligroso, very dangerous. None of the Shuara (Jivaro) wants to go there." Adriano asked about the caves in the area and his reply was that there are two caves and that both were on the Oriente side. He quickly pointed out that the caves were dangerous because they were home to the tigres during the day.

The Necta clan, whether because of contempt of civilization or excitement for our project, decided to cut their stay in Cuenca short and leave that afternoon for Sucua. Before leaving they thanked everyone for our hospitality, accepted a tent, an AR-7 and 100 rounds of 22 magnum shells. Antonio agreed to leave on November 26th with six Jivaro cargo carriers and meet up with us on December 3rd at his house on the outskirts of Sucua. He reminded us that the Rio Ojal was a dangerous place to go without heavy firepower. It would be cold at night and the Ojal had no leaves for building a fire. However, we could fish and kill monkeys to supplement our food supply.

"We must go in December and January after that time the journey will be too difficult and too dangerous," he said.

Oriolfo escorted them to the airport, Adriano left for the university, and I returned to taping my diary. Within the hour Oriolfo was back at the Vintimilla residence with the Jivaro. Their flight had been delayed due to bad weather in the Oriente. I was beginning to believe that there was no rainy or dry season. It seemed to be a perpetual rainy season with a few hours of sunlight.

Esus decided to take Maria window-shopping, leaving Antonio and I alone together. Antonio sensed my uneasiness due to my limited Spanish and spoke in slow and understandable phrases. He was curious and wanted to know more about my weapons. Why were the 22 magnums more powerful than the standard 22 shells sold in Ecuador? What was the most powerful American weapon for the tigre? I asked him to follow me upstairs to my room and began showing him my arsenal of firepower. His eyes grew wider and wider as he examined the 357 magnum and the large bullets in its chamber. Pulling out the Gun Digest that I packed around, I showed him the Magnum 44 rifle that would be in Cuenca within the week.

He could not contain himself, "Grandisimo," he whined in his high-pitched voice. We poured through the Gun Digest like two kids going through a Christmas wish book.

He told me that he owned an old muzzleloader. I thumbed to the section on shotguns. When we came to a photo of a Remington automatic, I told him that this weapon would be available for our expedition.

"This is my weapon of choice in la selva, (jungle) for the tigers," I said, as he looked up from the magazine and smiled.

Prior to Antonio's arrival in Cuenca, Adriano, against my better judgment, insisted we test the snakebite serum brought down earlier by Jim Bell. Adriano passed with flying colors but my skin had turned beet red and after nearly passing out, the medical intern at the local clinic told me that I was suffering from an acute allergic reaction to the serum. "Oh, this is just great," I said to myself. Then I read the fine-print instructions on the box that required the serum to be refrigerated.

"Terrific!" Adriano roared with laughter when I told him that if I were bitten in the lobby of a hospital, I might have a chance at survival. He retaliated by saying that at least he would have a chance. Sometimes Adriano's humor left me cold, but it was no secret that I was having serious problems with the jungle snake stories. I had researched snakes of the Amazon Basin and like a moth approaching the flame I was consumed by curiosity.

"Tell me about the culebras in the Oriente," I asked Antonio.

"Snakes come in all sizes and colors in the Oriente," he said. They were small and large and beautiful and ugly. Their colors ranged from black, blue, green, olive green, yellow, red, orange, brown, and chocolate brown, earthy, gray, plumb and white. There were multi-colored snakes like the deadly Naca Naca, which came in varieties of yellow, white, black and red, black and white and grow up to six feet and were deadly.

There was the Loro-machacuy, which was blue, white and black and green. They were up to eight feet in length, thin, and moved with great swiftness and some were tree climbers. A quick moving slender snake that measured up to nine feet in length, the Aguaje-machauy was orange to reddish chocolate, long fanged and after striking it coils and waits for its victim to die. The Mantona-venenosos, an orange snake, sometimes yellow, and white, could grow up to nine feet. Antonio was adamant when he said this killer would follow a man and strike without provocation.

But, to the Jivaro these culebras (snakes) were tolerable when compared to the fierce and deadly Shushupe. There were at least three varieties of the Shushupe, a pit viper Bushmaster. The deadly bite from the one and one half inch fangs was fatal in five minutes. This terrible snake lived in holes in the ground and its coloration was reddish to yellow, sometimes pink with black bands. Very alert and aggressive it followed a man and killed either by day or night. Antonio said that the Indians fear this snake above all others because it did not strike once and back off and wait. It continually struck until the victim was dead. When he spoke of the Shushupe, Antonio's eyes became wide and his voice very high pitched. "Cinco minutos," he said. Within five minutes its victim died.

My God, I thought to myself, do I really want to pursue this conversation? After a brief hesitation of stomach churning, I cast aside my uneasiness because of the invaluable information that could help me understand what we were up against. I pressed on.

"What snakes do you fear the most?" I asked.

Without hesitation he replied, "The Shushupe, Tres Minutos, Coralito, Dos Minutos and the Equis."

The Tres Minutos, or three-minute snake that Antonio referred to, was probably the Fer-de-lance, one of the most terrible snakes inhabiting Jivaro territory and the Amazon. The Fer-de-lance is not only a relative to the feared Shushupe Bushmaster, but was also related to the poisonous tree vipers. Thick bodied and flat-headed, it had catlike eyes that contained a pupil. It was slower than the others, but struck faster from a coiled position. Its long, poison-filled fangs injected a great amount of venom, which attacked the blood system, causing internal hemorrhaging. A deep pit between eye and nostril was thought to be the organ of some unknown special sense.

Antonio said, "It has a sixth sense and knows when man is coming into his domain." Behind its lance-shaped head the rough-scaled seven foot body was gray crossed by dark bands margined with yellow or green. This nocturnal snake was viviparous, and had 30 to 36 young in a litter. When it swallowed its prey it dislocated each side of its jawbone. And it attacked without provocation. According to Antonio it was especially dangerous when surprised or during mating season.

"The machete or a scatter gun is your best defense against its attack, if you have the time to defend yourself," he said matter-of-factly.

The Coralito, or the Coral snake, is related to the deadly African Mamba and the Indian Cobra. It has a red coral-color rings on a brown to black body, could grow up to four feet in length, and had a small blunt head. Due to short fangs the Coralito had developed a peculiar side-swinging strike. Then it hung on after piercing the flesh and began a quick chewing motion. Death is caused by shock to the nervous system. The poor victim died from lack of breath and in extreme pain.

The Dos Minutos was probably an unclassified tree or pit viper and named by the Jivaro, because its victim usually died within two minutes. "Perhaps this is the Equis?" I asked.

Antonio quickly dispelled this idea. "The Equis is muy peligroso. It is small compared to the nine-foot stalkers. Equis is between six inches and several feet. It attacks a man because it has a bad mood. When it strikes it will jump a distance twice its length and the bite is deadly. If the Equis bites you on the finger or toe, you must cut them off to save your life," Antonio said. "Many of my brothers in the Oriente are missing fingers from their hands because of the Equis."

How in the world could a man take his machete out of its sheaf, place his index finger on a rock, and with one blow severe it from the body? Would I be willing to make this sacrifice? Could I trade a body part for my life? The answer was not a simple one. Reassuring me, Antonio grinned and said that he had often wondered what choice he would make.

Seeing my visible uneasiness with the snakes of the Amazon basin, he told me not to worry because the Jivaro had secret cures for many forms of snakebite. Antonio said that he would go deep into the jungle and cut some "Piri Piri" for me. "It will protect you from some of the snakes if you are bitten," he said.

He was now at ease with me and both of us recognized that a friendship and trust was beginning to grow, so I felt comfortable in opening the next topic of conversation. Esus had previously given us a tiny bit of insight into Antonio when he mentioned in passing that Antonio had helped an American with a disease by providing him with a "juice" from the Ojal region. "Tell me about the juice that you gave to the American doctor", I inquired.

He said, the "juice" was actually the bark from the Chuchuhuasha tree and when pulverized, mixed in alcohol and taken orally cured cancer. The American doctor had sought out Antonio for his research into medicinal plants. Antonio had taken the American up the north bank near the Ojal and supplied him with the bark and other herbs and plants, but when they tried to cross the Tutanangosa, most of their supplies were lost. Shortly thereafter the expedition was cancelled.

"There are many Shuara cures for diseases," he said. His pharmaceutical warehouse, supplied by Mother Nature, included an herb from the Ayamullaca for curing open wounds, a tree-bark juice from another tree that was used as a blood purifier, and a vine named Paushi that contained juice for tuberculosis. He described a giant grass similar to bamboo that contained water, which was used by the Jivaro to cure liver ailments.

His jungle warehouse also included aphrodisiacs, vitamins, cavity deterrents, and snake serums, perfumes, weight depressants, antiseptics and other cures. The Ucho Sanango tree was a well-kept secret and the keeper of the knowledge belonged to the Acuara Shuara and the Morona tribes located deep in the Amazon. The leaves were boiled together with some leaves from the Renaquillo tree and taken orally by those suffering with broken bones. He said bones completely healed within seven to fourteen days.

When asked how he obtained this information he replied, "I must take long trips into the jungle to visit with the uwisin of other Shuara tribes that have this knowledge. In order to receive knowledge, I must give knowledge. Uwisin (shaman) power and respect resided in trading knowledge for knowledge."

"Are you an uwisin?" I asked

"Si Lee, I am an uwisin," he replied.

He explained to the best of my understanding how his decision to combine the forces of nature with the power of the supernatural went against the grain of pure tribal shamanism and forced him to walk a fine line in the Shuara community that he served. I assumed that in order to appear different than the normal uwisin, he adopted western clothing and embraced the religious leanings of the Salesian Order while at the same time he obtained plants from the jungle to assist in his respected practice as an uwisin.

Antonio explained that uwisins (shaman) are of two types: bewitchers and curers. Necta was a curer, a pener. He smiled, remembering his first curing when he treated a young girl for snakebite. Her family believed that the attack was brought on by a wawek, a bewitching uwisin. Necta was given a cow for his healing services. A single treatment to remove a wawek curse could net an uwisin a shotgun, blowgun or cow.

Understandably, Antonio's power as an uwisin, called Kakarma, increased his intelligence, physical strength, and made it difficult, if not impossible, for the uwisin curer to lie or commit dishonorable acts. His acquired power, according to my research into Jivaro tribal legend, also increased his resistance to contagious diseases and made it difficult for a bewitcher (wawek) to harm him through physical violence or sorcery. Thus the uwisin curer, as long as he maintained his power, could roam freely with the spirit helpers throughout the forests without fear.

The spiritual servants of the Jivaro, commonly known as magical darts or tsentsak, were most powerful for a curer. The curer or pener uwisin could cure or bewitch, whereas the bewitcher or wawek could only bewitch. However, Necta said the power of the tsentsak was gradually used up through curing or dispensing and he was careful to be discerning of its use. Thus, as a young man he decided to embark on a new adventure, which would allow him to preserve his kakarma by utilizing nature's cures. In time and with support from the tribal elders he learned the medicinal secrets of nature found in the womb of the Macas Forrest.

It became clear to me that he had not found the emeralds of Mejia, as Alfredo had suspected, nor had he inherited wealth, as Adriano had surmised. Instead, he had earned his wealth by protecting and curing his tribal members through his knowledge as a curer. Through his sensitivity and talent he became an important asset to the welfare of his tribe and for years they had sought him out to administer health and protection. His hard work and dedication made him wealthy in material goods by Jivaro standards. He seemed proud of these accomplishments that brought him two homes, cattle and vaqueros, cowboys to work his beloved Wakani ranch. His family never went hungry, had a comfortable home life and, most importantly, they were respected in the settlement of Sucua. He said this was the life he dreamed of as a young boy when he made the decision to become a pener uwisin…a curer.

Then he asked if I believed in spirits and added, "My people believe in spirits. We believe there are both good and bad. I am a pener uwisin that works with good spirits and they told me to come here and meet with you".

Before I could react, Oriolfo appeared with Esus and Maria and said that the plane would be leaving within the hour and that they should find a taxi and get to the airport as soon as possible. Saying goodbye to my newfound friends, I couldn't help but admire these fearless people that lived in danger with the shadow of death constantly peering over their shoulder. One nagging question remained locked in my mind. Why did he ask if I believed in spirits? This was the second time this week I had been asked that curious question.

"The National Liberation Front inflicted heavy casualties on the American Imperialist forces this weekend and will continue to do so until the lackeys of Kissinger and Nixon come to their senses and withdraw from Vietnam." Saturday night in Cuenca and I was in bed at 10:30 listening to Havana Anna on Radio Havana, killing time until I left for Guayaquil to meet up with Bell and new blood. I spun the dial on the radio. Radio Quito just reported that the Aucas in the Rio Napo region of Eastern Ecuador killed twenty-eight people yesterday. I thought of Connelly. The news was so depressing that I switched back to The Voice of America and began to clear my mind with Mozart.

Jim's mention of new blood was intriguing. Who in the hell—and in their right mind—would be willing to come down here? I didn't much care whoever it was, as long as they spoke English and had a sense of humor; we would get along real well. In three days Adriano was scheduled to take his final exams for his doctorate of law, Jim Bell would arrive with funds to set up the Guayas Mining Company and Antonio and friends would be headed up the north bank. It was amazing how the law of timing works. It seemed like you waited forever for things to happen and then suddenly everything comes together at once.

The Saeta Airlines turbo prop skimmed over the massive cemetery stretched out on the mountainside below, banked to the right and landed on the sun-bleached runway of the International airport in Guayaquil. It was always the same when you flew into the Pearl of the Pacific. The first landmark seen from the air was always "the muddy and fever-haunted Guayas River", as Colonel Fawcett labeled it years earlier. The second was potters field that ran up the side of a small hill where the poor and downtrodden of this teeming city were laid to rest on top of each other in simple graves marked with white crosses.

The clean freshness of Cuenca was soon a distant memory. Another cab driver with a death wish navigated the bustling, diesel smoked streets in a mad dash to the Humboldt Hotel. It was always an inward struggle to return to this place after experiencing the tranquility of the mountain fortress of the Incas. Cuenca had its charm, but Guayaquil had its lure of good food and beautiful women, who were eager and willing, providing you, had a car and an apartment for a secret rendezvous. Unfortunately, I had neither and it was an impossible dream to try to sneak a beautiful senorita into a hotel room under the watchful eyes of the stringent hotel staff and Sr. Ortiz their commandant.

Sr. Ortiz informed me that Senor Bell and a friend had checked into the hotel at 6 a.m. Both looked tired from their long flight and struggled with heavy luggage that required the entire bellhop staff to carry it to their room. Apparently one of the duffel bags was so heavy that it almost jerked one of the staff workers arms out of socket when he tried to lift it. "Just mining equipment", I said before he asked.

It was no exaggeration to say that I was ecstatic, yet saddened, to learn that the new blood was my brother-in-law, Bill Wright. He was certainly qualified for jungle duty, an ex-Marine with jungle and weapons training and a physique that had been chiseled by many years of construction work. Bill was also the type of guy that you wanted in your corner when things got rough. His motto, "When things get tough, the tough get going." Bill had convinced Larry Webb and Jim Bell to bring him down to be my point man. Bill, a good old boy from the deep south, raised somewhere around Lookout Mountain in Georgia, stood about five-foot-ten, and sported a perpetual deep tan that enhanced his sun bleached hair. When needed he had a southern drawl that could melt cold butter.

Bill was an ideal selection but my emotions were divided. I knew that my sister, Betty, and their three kids, Bryan, Brendan and Bambi Lynn were totally dependent upon Bill for their economic welfare and survival. The Mejia project would certainly not qualify as economic security. It was a long shot at best, a gamble with life and limb against insurmountable odds. But Bill was three-times-seven and willing to risk the odds that confronted us and he was here. I let my concern go and welcomed him. It was also good to see Jim Bell and his easy going smile.

"Hoss, the Tigers won the World Series," Bill said with excitement. "It started off a little rocky for them when Bob Gibson pitched a five-hit shutout and the Cardinals won the first game 4 to 0. The series went seven games. The Tigers came through in game seven to win it all. Kaline, Cash, Horton and the rest of the guys sucked it up and now are world champions. In fact the experts say that this World Series probably saved Detroit from heavy-duty race riots. Can you imagine racial violence taking a back seat to a sporting event?"

Bill was animated and continued, "Hoss", which is what he called me, "you wanted more firepower? Look at what we brought down." He unzipped the drab, olive green duffel bag and began removing guns and ammo. "This little jewel will stop a big cat in its tracks. It's a sawed-off Remington shotgun with a double aught load. This is the weapon of choice by the grunts in Nam for jungle fighting at close range. If an attack comes it'll be sudden and in thick jungle. We'll only have one chance and one round to get the job done. I sawed it off so we have a larger kill radius and with the double aughts, the cat is in a world of hurt. He won't have time to maim, only die."

Bill continued rummaging through the weapons cache and I got the distinct feeling that a drill Sergeant was prepping me for my first search and destroy mission.

He continued, "Now, if a cat decides to dog our trail and won't come in close enough for our shotgun, then this little old 44 magnum carbine will take him out over a hundred yards away. We brought three hundred rounds for each of these babies. Here's three more AR-7s and five hundred rounds of 22 mags.

"And look at this Hoss! If the shotgun jams and I get a second chance then this pistol is the equalizer."

It looked like a freaking canon! My eyes lit up with joy while staring at the lethal firepower that would finally place our expedition into the womb of the Macas Forest on a level playing field. "Antonio will be delighted," I said while caressing the 44-magnum carbine.

Bill added, "I brought you ten jars of Peter Pan peanut butter. And, we have new stainless steel machetes, canteens, Cutter Snakebite kits, backpacks, bandoleers and a lot of WD-40."

I didn't dare ask how they managed to bring the weapons in under custom's radar. I really didn't want to know, but I had the feeling they were about to get into it, so I changed the subject and told them that Don Antonio Necta would be headed toward the Rio Ojal at first light.

CHAPTER 9

UWISIN

Lee Elders in Sucua

Necta and five of his Shuara brethren left the sanctuary of Wakani before dawn and headed due north, up the bank of the Tutanangosa. He had memorized our instructions and made notes regarding Mejia's riddle and remained confident that the rock open in the middle could be located somewhere along the bank of their destination, the Rio Santa Rosa. He hoped the advance expedition might produce good results providing, of course, that the weather remained favorable. When he scanned the western horizon not a single puff of white cloud was seen. Good weather was holding.

He said their first rest break came when they stopped to survey the swollen bank of a small stream. After a smoke they found a shallow spot to wade across. When they reached the other bank Necta felt pressed to deploy camp and leave his fellow Shuara for a one-day journey in search of the piri-piri. He knew my concern about the snakes in the region and felt obligated to deliver the snakebite cure by the time we joined them. He bid farewell to his men and headed into the bush, armed with knowledge, a machete, my AR-7 and 50 rounds of hollow-point ammo.

His private journey took him in a northerly direction where he searched the highlands for the Almendro tree that was used for a type of medicinal oil. He knew that if luck smiled on him, he might be able to locate the Catahua tree and extract its juices, which limited infection and caused any poison to recede to the original point. The magical Catahua juice was often used after battles to heal wounds from arrows or spears that had been dipped in snake venom. The Shuara believed the narcotic resin was also an effective cure for leprosy, a "white man" disease brought into the Oriente by the colonials. Recently Necta had learned that a fellow uwisin, who lived several days' march to the north, had been claiming these cures. But his main objective this day was to honor a commitment to me and find the Piri-Piri, a tuberous root from the plant that not only cured some forms of poisonous snakebite, but was also used as a perfume by some tribes and an aphrodisiac by others.

Necta was indeed an enigma. He embraced the traditions of his fellow Shuara, the name preferred over Jivaro, which was the identity given to them by the western world. Yet, years earlier he had forgone his native dress in favor of the more practical clothes of the colonials which included pants, shirt and rubber boots that made walking in difficult terrain more comfortable. When crossing rivers and wading through parasite infected mud the rubber irrigation style boot was a godsend. He had long ago abandoned the typical estemat headdress, a woven cotton band with toucan feather tassels, in favor of an American style baseball cap emblazoned with the large, yellow letters CAT.

Necta seemed torn between these two opposing worlds that vied for his heart and soul. His allegiance was divided between the peace and prosperity that was seemingly at the core of the colonial's civilization and their religion, which contradicted his traditional Shuara knowledge. His traditional side included bitter inter-tribal warfare that often resulted in pain and bloodshed, which he found pointless. He understood the Catholic purge of the ungodly act of head hunting, which had been the sacred practice of the Jivaro. Even the Shuara considered trafficking of the tsantsa, the shrunken head trophies, sacrilegious. The tribal ritual had originally been a means of obtaining power to those in the interior.

I had made mental notes of how Necta marveled at the modern day appliances and technology that were finding their way into Sucua. An airport and planned road link between Macas and Mendez and bilingual Jivaro schoolteachers in remote locations were winning over the hard-liners of his tribe. Many of the Shuara believed this would provide a better life for their children and loved ones. But, as a Shuara in the white mans world, he grieved to see civilization gobbling up their land at a rapid pace. Colonials were moving into the jungle neighborhood. The Church had bought up land on the south bank of the Tutanangosa and placed the land off limits to everyone, including his tribe.

He seemed to be a man of mixed emotions and mixed traditions, who carried a mixture of Spanish and Shuara blood and a name that paid homage to both: Antonio, the Spanish and Necta, the Shuara. Every aspect of his life was trapped in a personal dilemma of a man torn between two ideologies influenced by distinctly opposing cultures.

He had admitted that by design he had bridged both worlds with honor and respect. He seemed to have a knowing about him, a natural instinct that produced confidence, yet he understood the power of the uwisin was constantly changing. The ebb and flow of survival was fluid and a wawek with powerful magic darts might be able to penetrate his delicate defenses. His balance came from a good rifle with the white man's ammunition because he knew that this might be his edge if push came to shove. The thought comforted him as he headed deeper into tigre country in search of the elusive Piri-Piri.

Sometime around midday, with the sun shining directly overhead he spotted the tall grassy clearing, which signaled his arrival in the fields of fresh Piri-Piri. Pulling a knapsack from his backpack and his machete from its sheath he began digging into the root of the plant, removing the bulb and placing it into the sack. The end of the root bulb was the source of power that made the piri-piri effective against snakebite. Once he determined that he had an ample supply of the magical root, he headed into a thicket of strange, exotic trees that provided more magical medicines for his kit. Slicing the bark to capture small amounts of the lifeblood of the trees, he blessed the magnificent juices that provided relief to so many. He carefully packed Mother Nature's remedies in his sack.

Alone in the womb of the Macas forest he felt more at ease with his native heritage. His comfort zone was indeed the jungle, a mysterious place that encompassed beauty, power, danger and raw nature. He believed that if a man could harmonize with the jungle then the polarity between good and evil could be neutralized and fear would not rear its ugly head.

Deep in thought, and in control of the natural forces surrounding him, he wondered why the white man was so afraid of this amazing place. Why did they only come here in search of tiger skins, gold or treasures when they could learn so much about themselves? He questioned whether they would learn that the Amazon rain forest is nature's pharmaceutical house and had cures for man's illnesses and diseases? His dark eyes sparkled with life as he posed questions that didn't require a verbal answer. He was in a place that was rich in the essence of nature, flourishing with exotic trees and the jungus plant where the piri-piri and his beautiful childhood memories could be found.

The magic of the area had lulled him into lingering beyond his allocated schedule. The sun's rays were casting slivered shadows through the canopied jungle. He knew that he must hurry to get back to camp before darkness set in; even an uwisin respects the jungle at night.

Necta later admited he had misjudged his timing and nightfall had set in making his surroundings more foreboding than ever. Feeling part of the jungle and its shadowy natural rhythm, he stood on a small ridge listening to the rumble of the cascading Tutanangosa and knew that camp was in close proximity. But which way would he go? Did he head west up river or east down the river? A dim glow from a campfire to the east caught his attention and he knew immediately that was his beacon to safety.

Before long he strolled into the Shuara camp, receiving saludos from the men he had left behind. He felt good to be among friends and near the warmth of the fire when the local brew was passed around freely. Manioc beer was as important to a Shuara expedition as a good steel machete. The Canari loved their aguardiente and the Shuara enjoyed their manioc. Manioc is tuber usually grown in a garden, harvested, cut up, and placed in a pot to boil. After boiling the tuber is then mashed and stirred usually by one of the wives. Stirring the mash entails more than the swirl of a spoon in a pot. As the woman stirs the mash, she chews handfuls and spits them back into the pot. The mastication of the mash is considered the aging process for the rapid fermentation of the potent brew. Legend had it that the beer is best when chewed by a pretty girl rather than an old woman and Necta's beer would qualify as one of the best as it had been prepared by his beautiful young second wife. Songs, storytelling and beer drinking helped ease any fears brought on by the jungle, which was centuries away from the colonial city of Cuenca.

Adriano Vintimilla Ordonez sat on a straight hardback chair in the center of a cold and drab classroom. He answered numerous legal questions from a group of seven Professors of Law from the University of Cuenca. It was his final oral exam to obtain his Doctorate of Law degree and one of the most important evenings of his life. The prestigious event was open to his family and friends and we all sat in the back of the room silently rooting him on. Flanked, on both sides by his peer group, his composure never once waivered as he calmly answered the difficult questions that covered everything from domestic to criminal to constitutional law. While he gave the performance of a lifetime, my mind raced back to that morning over breakfast when he confided that one of the instructors, a communist, would surely give him a failing grade.

During his senior year, Adriano had become active in university politics and was elected class President of the Christian Democratic Party, which was a proponent of the American envolment in Vietnam. His formidable opponent of the war was a university professor and self proclaimed communist who attacked the C.D.P. at every opportunity. They had many spirited and heated debates on the Vietnam subject and it wasn't until mid-term that Adriano discovered that his adversary would be one of the professors participating in his final exam.

As the evening wore on Adriano's family and fiancé swelled with pride and confidence, certain that his hard work would be honored with the doctorate degree. I sat quietly; my stomach churned with an uneasy feeling that politics might undermine fair play. Finally, after three hours the inquisition was over and we retired to the university lounge to await the results. When the verdict came in my concerns proved unwarranted. Adriano Vintimilla Ordonez was formally a Doctor of Law and to punctuate his hard work, he had received the highest grade ever given by the University of Cuenca for an oral exam.

The Cuenca Tennis Club was jammed with over two hundred people milling around with drink and food while awaiting the entrance of the guest of honor, his family, and friends. The guest list included state senators, judges, university professors, the Mayor of Cuenca, and too many of the Vintimilla clan to count with ages that ranged from babies to eighty. Families with their children, who were playful and loud, filled one corner of the large ballroom. Another section was reserved for the elderly and another for the single women that huddled together, smiling and whispering among themselves each time a handsome male strolled by. Forty bottles of champaign chilled on ice while forty bottles of Johnny Walker Red tempted the heavy drinkers from the shelves behind the makeshift bar.

Everyone was dressed in their finest. The women were attired in flowing gowns purchased in Madrid and Paris and accented by gold from the Oriente that had been fashioned into jewelry by Cuenca's renowned artisans. The men, clad in black suits, starched white shirts, black ties and shining black shoes, gathered in all male groups to sip their scotch and discuss politics, world events and the continuing depreciation of the sucre, their unpredictable currency. Everyone was dressed for the occasion...except the three of us Americans. Our closest suit was four thousand miles away.

Inside the club we headed straight for the bar so that we could remove ourselves from the attention we were getting and plan our strategy for the evening. Unfortunately, the route to the bar took us across the center of the dance floor and we couldn't have been more pretentious had we asked for a spotlight to accompany us. We nervously sidled up to the bar and ordered three double scotch and waters and tried to relax.

A loud drum roll silenced the crowd. The master of ceremonies introduced Adriano and Sonia, touting the new layer's final exam score as the highest ever achieved, and informed my buddy that he was Cuenca's newest member of the exclusive Doctor of Law club. Almost as an afterthought he added that Adriano and Sonia would be getting married in the near future which brought almost as many cheers as his entrance to the law club. In honor of their son, Adriano's parents were brought to the center of the ballroom to commence the first dance. Manuel Vintimilla Crespo, dignified and ramrod straight, led his beloved Virginia Vintimilla Ordonez in a spirited dance that brought down the house.

The party went on until the wee hours of the morning and when we finally got to the hotel, the doors were locked. Fueled by scotch we went a little berserk and woke up the entire neighborhood bellowing the famous military beer drinking song. "99 bottle of beer on the wall, 99 bottles of beer, you knock one off and what do you have 98 bottles of beer on the wall, 98 bottles of beer, you knock one off and what do you have 97 bottles of beer..." We got down to 78 bottles of beer on the wall when the night desk clerk finally unlocked the door. He had a few choice words for us as we staggered up stairs to our room and fell into our swimming beds.

The Shuara hunting party struck camp at sunrise and headed due west. The next major river they encountered was aptly named the Tigress. Beyond this small tributary lay the hunting grounds of the ferocious and magnificent jaguar that was both revered and feared by the Shuara tribes. Armed with only two twelve gauge shotguns and the AR-7 they crossed the Tigress and immediately were snared by the thick undergrowth of the Cana Brava, a palm cane that grew in thickets, sometimes miles wide, and was used by the tribes for ceilings in their houses. It was a terrible place and Necta instructed everyone to stay close and be on full alert. Stragglers would be too tempting for a hungry predator who might stalk and attack from behind.

Big cats preferred to launch their attack from height; utilizing their weight to break the back of their prey on impact. Once the prey was down they crushed the skull between their jaws. Their powerful jaws could crush through a sea turtle shell, so a man's skull didn't stand a chance. Every now and then, Necta would signal everyone to stand still and be quiet as he listened for grunting moans, the telltale sign that a jaguar was in the vicinity. When no sounds were heard they cautiously moved forward through the swaying grass. Once out of the Cana Brava thicket they found themselves wading through pale green foliage slightly bending in the morning breeze. Near a clearing they decided to build their camp for the night.

Necta felt something strange in the air. It was a feeling that only an uwisin would discern and he knew that he must act quickly. He located a thicket of durable, black, Chonta wood used to make arrows, bodecaros (blow guns) and lances. Quickly it was cut in ten-foot increments and positioned around the perimeter of their sleeping area. They shaped and sharpened the tops into pointy spears and bound the poles together with Marona palm cane shard. The blunt end was buried in the ground and slowly their ready-made fort, supported by large rocks began to take shape. It was just a hunch but he felt that their base camp needed protection.

Four strong men constructed the fort; another prepared the evening meal and brew and readied the one tent brought from Cuenca. The tent was solely used to keep the provisions dry from the fog and mist that rolled in every evening from the waters of the Tutanangosa, a stone's throw away. Firewood was stacked in the corner of the fort because during the night their most valuable asset was a roaring fire to keep them warm and frighten off any wandering predators. Fire attracted then repelled night predators so they slept in shifts to make sure the fire did not die out during the night. Completing their tasks they sealed the narrow makeshift gate and bedded down for a long and semi-sleepless night.

Necta told me later, that sometime after midnight when the full moon was high the first signs of danger came in the form of the jaguar's throaty cough that seemed to ring the fort from several directions. The fireguard, Yaupi, yelled an alarm to the sleeping men and bedrolls quickly emptied as everyone scurried for their machetes and weapons. The Shuara knew the niawas (jaguars) were hungry and hunting in packs. Quickly, Yaupi fired off one round from his muzzleloader for effect. The noise from the old blunderbuss was deafening and startled everyone, including the predators.

The men stood as statues and listened for signs over the rushing waters of the river. The gutteral coughs of hungry niawas were heard from all directions. Something was wrong. Necta knew the niawas were solitary predators. Why were they hunting in a pack? Necta closed his eyes, feeling into night and the jungle. Yaupi started toward the fire: Antonio raised a hand. All stood motionless in silence. The niawas had not left. They had regrouped and were milling around their camp. Necta ordered everyone to fire their weapons and shout at the top of their lungs hoping to dissuade the predators from their intent. The noise generated from the hail of gunfire seemed to work; the niawas retreated. He said it was a long and sleepless night for the battle-ready men.

Necta expressed his concern because Jaguars were territorial and they mixed together only during mating season. At any other time they were considered loners. Why, on this evening had they banded together in force to approach his campsite? He admitted later he was worried. Was he under attack by a wawek utilizing powerful tsentsak? Or was this evil a common occurrence when camped in the foreboding womb of the Macas Forest? This was a place where many brave Shuara refused to go. He had heard of marauding niawas from others that had survived to tell their stories. Perhaps he was overreacting, but he knew he had to rely on his powers as an uwisin and his connection to the jungle.

The following morning after little sleep they left in single file for the Rio Santa Rosa. They errected base camp and one man stayed behind to protect the supplies. The rest of the men worked their way up to the Rio Santa Rosa and attempted to judge the height of the mountains around them. The only way to search for visible clues left by Mejia in the maze of green was to take to high ground. The cliffs across the Santa Rosa seemed to be a good starting point. To reach the cliffs the group had to wade across the fast flowing river. A daunting challenge because of slippery moss covered rocks; one mistake could cause a man and his supplies to go tumbling into an icy watery grave. Caution was the byword as they struggled, single file trying to stay upright while using their wayacan walking sticks for balance.

Once they safely crossed the bank of the Santa Rosa the team fanned out, searching for a trail that would lead them upward to a visual vantage point. A deep rut carved by water runoff served as a trail of necessity rather than choice. The climb took several hours but they took their time, always on the lookout for dangerous tree vipers that could swing from an elevated branch and strike them in the face or upper body. A strike in this vulnerable region of the body meant a painful death, as the venom injected would travel quickly to the heart.

When they reached the crest they scanned the massive green tapestry of the wondrous rain forest that stretched before them. They were amazed at the number of derumbes (landslides) that dotted the land. It was evident to Necta that this Mejia clue was of no use to modern day explorers. In eighty years things change and places are hidden by the constant growth of the jungle. The only recognizable things were the tributaries below that rolled and crashed into the defiant big river, the Tutanangosa.

Necta spread out his old worn map on the dry ground and began to double-check the rivers they had already crossed. They were definitely on the mysterious river the Colombians had named the Santa Rosa. The rock that was open in the middle was Antonio's next objective.

"The mineral stone is in the Rio Santa Rosa." Necta surmised that if the Colombian was truthful in his final instructions to his family, then he and his men would not have to cross the Santa Rosa. They waded back across the river and stood on the Oriente (East) bank, at the confluence of the Rio Santa Rosa and the Tutanangosa. Taking out his notes from the meeting in Cuenca meeting he read:

"From the shore to the Oriente side, walk until you find a landslide, from the landslide four-hundred meters down, a large stone open in the middle. The emeralds, tres arrobas, covered with other rocks in this stone. In front of this stone, a half a cuadra (city block) runs the water of the same river."

Before striking out up river on the Santa Rosa, Necta gathered his men together and instructed them on what to search for. He split his team into two groups. One group would work their way up the east bank and the other would begin their exploration a half a cuadra or about fifty yards inland. With one team searching inland and the other searching the riverbank they should find the "large stone open in the middle."

Necta's plan was sound until his team was upstream approximately one hundred meters and found they could no longer follow the meandering river. A stand of fallen tree trunks, stacked in crazy-quilt formations, blocked their path. If they tried to work their way through the fallen timber they faced unknown dangers. They had been lucky so far avoiding encounters with fanged killers, but were fully aware that their fortunes could change at any given moment. He and Yaupi climbed up a leaning tree trunk to get a better view of what lay ahead.

Suddenly, Yaupi grabbed Necta's shirtsleeve. Yaupi's primitive instincts detected the ever so slight movement that betrayed the hiding place of a vicious yguana machuacuy. The brown and black yguana blended in with the bark and limbs from the fallen tree. The ugly snake with the iguana-type head was six feet of sluggishness that preferred to lie in wait for its prey. Terror was waiting silently and patiently for its unsuspecting breakfast. Non-aggressive the serpent was still deadly if its victim could not find aid from an Indian brujo. Piri-Piri could neutralize the venom, but Necta decided to backtrack rather than attempt to crawl over and under large logs that were the perfect nesting place for other reptiles searching for rodents and insects.

North of Necta, the inland team wasn't having any better luck. They found themselves cut-off from forward progress by a steep incline. They were forced to retreat and try to maintain, a distance of fifty meters from the river. A difficult task for the Shuara, who were preoccupied with their own safety while trying to blaze a trail through the maze of thick jungle in search of a large rock open in the middle.

Necta said that the two teams had accidentally come together in a narrow ravine that circumvented the steep ridge and fallen logs. The tired and sweaty men were overjoyed to see each other and they exchanged stories of their difficult circumstances and decided it was time to head back to camp. Now that they had found a narrow doorway up the river they planned an early start the following day.

They spent most of the next day chopping a trail Necta estimated to be seven hundred meters from camp. They were bursting to relieve themselves, but they knew this would attract predators to their trail, so they decided to chop a new trail down to the river. The river water would dissipate the ammonia smell from their urine and also carry their waste downstream away from curious predators. It took a half-hour to accomplish the task, but to the Shuara who were adept at surviving in harsh conditions, it was wiser to be safe.

Nightfall rapidly descended on their tiny fortress where a roaring fire cast an eerie glow in the surrounding forest. The men drank their daily ration of manioc beer, which in their native tongue was called nihamanci and conversed in soft tones about their lives, loves and the day's hardships. Necta admitted to me later that he felt restless and withdrawn that evening. Something about the Mejia riddle was bothering him. Pulling the handwritten notes from his backpack, he moved in close to the fire and began reading and re-reading the mysterious passage that defied discovery for so many years. *"Like making a relation, the entrance is by the cascabel..."*

He remembered that during our meeting in Cuenca he had quickly identified the cascabel as a tree that grows wild in the Oriente. He felt embarrassed by his statement because this tree was everywhere, which meant it was not the cascabel Mejia had spoke of. It had to be something else, he thought, but what? Was it a landmark of some sort, like a tunnel or cave? Perhaps it was a waterfall? A waterfall makes a crashing sound, a rattling sound, when the water falls upon the rocks below. But what was the entrance? Was the entrance a trail that led to the outcropping? Or was it the entrance to a tunnel or an old Inca mine? What does, "like making a relation mean"? This part of the puzzle bewildered him as he tried in vain to make sense of the riddle before surrendering to sleep.

The riddle became the least of his concerns. Shortly after midnight they heard the first whistling sound above the crackling of the red-hot fire. The Shuara remained still and listened, cocking their heads from side to side as the whistling sound seemed to be coming from the trees and undergrowth around the fort.

"Culebras," Necta shouted above the din of shrill sounds that were cascading down upon them. The Shuara grabbed their makeshift torches and held them above their heads, straining their eyes in the dim light for movement from the phantom whistlers that seemed to have encircled them.

They could see leaves and tree branches swaying from the movement of tree-hopping reptiles that signaled each other with eerie sounds like a high-pitched flute. They stood motionless in the glow of their torches, their eyes searching for any movement within the compound that signaled an unwanted intruder. They wrestled with the uneasiness swelling inside knowing that a frightened man emits an odor, an odor of fear, which snakes are sensitive to. The Shuara had been schooled at an early age that a snake's tongue moves in and out of its mouth trying to pick up whether their adversary was fearful and might hurt them. When fear exists in man, the reptile senses it and that was when it struck. The Shuara were also aware that when they believed in themselves or the power of an uwisin then no odor of fear was released.

The men knew the power that Necta possessed and placed their faith and trust in him while the strange serpent ritual continued. The men stood in a semi-circle unifying their inner-strength while protecting their backs with each other's presence. Not one snake breached their fortress. The Shuara acknowledged the mystical powers that protected them, but each was wondering if their ongoing danger was the powerful manipulation of a wawek, a shaman with evil intentions?

The parrot snake a tree-climbing verde loros was probably the source of the whistling sound, yet none of the Shuara had ever seen the whistling snakes up close or during daylight hours. The fact that they could not visually identify the reptiles made no difference to them; they knew from past experience that different species of snakes displayed different traits. The shushupe was cruel in its intentions and would mimic the faint sound of a duck that sounded distressed, as if in trouble or lost. The 12-foot reptile of immense cunning would diabolically lure curious victims and in minutes its paralyzing venom would cause death.

The night-hunting rattle-less jararaca, an orange colored diamond-headed fer-de-lance were as big around as a man's thigh. These night crawlers were stalkers and preferred aggressiveness, not deception to spew out their nasty venom that was painfully lethal. There were both noisy killers and silent ones that constantly searched for prey after dark.

Necta's team had placed their faith with him and for this he was grateful, but unknown to them, he later admitted he was bothered by the myriad of visous attacks that had occurred in such a short period of time. First the tigres rose up against them and then the culebras. The concentrated assaults were not only odd but also unexplainable. They went against the laws of nature and the norm of predators. He was now convinced it was the work of a wawek, a bewitcher with powerful magic, influencing and directing the predators of the forest against him. But who was he up against and why? His confidence in his power as a pener uwisin was under the scrutiny of a mysterious wawek and himself.

Necta said that the following morning he and his men left their fort and headed up the east bank of the river named after a saint. Groggy from the night's bizarre events they struggled under the weight of their backpacks, weapons, and machetes, making their way towards yesterday's stopping point. This would be their last day to search for the rock that was open in the middle because they were scheduled to meet us in Sucua in a few days. They knew that their progress was dependent on luck and sheer will power. Once again they divided into two groups and struck out.

By noon the second group, headed by Wajai, was making good progress along the low ridge when he and Ikiam stopped for a smoke break. Sunlight dancing against a large white rock below them caught their attention. They inched their way down the steep incline toward the rock formation, but their attention was again diverted, this time to a large cave that was partially obscured by the thick foliage. Had the sun not been reflecting off the rock they would have totally missed it.

Wajai decided that they must explore the cave and began cutting pushi vines to make a makeshift rope that he could use to lower himself into the subterranean entrance. Within a short period of time the two men had over forty feet of vine rope, which they anchored around a tree. Ikiam stayed topside to guide Wajai and the vine towards the destination point and act as a secondary anchor to keep the vine taut. Wajai slid out of his backpack, slung his shotgun around his right shoulder, and sheathed his steel machete. He was ready to make the drop.

Necta and Yaupi made good time until they found their progress blocked by a thick strand of hardwood trees that defied cutting or flanking. Forced to either backtrack or cross the river, they opted to cross the Santa Rosa to skirt the trouble area. They had to be careful; the river was a surge of swift and contrary currents with swirling foam and moss covered rocks disguising its true depth. The two men used a sturdy tree branch as a support but still struggled against the strong current of the chilling waters. Through sheer determination, coupled with a little luck, they reached the opposite bank and emptied their rubber boots of silt, small rocks and muddy brown water. They rubbed their feet and legs to slowly bring back circulation to their numb limbs.

Both men slumped against their backpacks and lit up some raw tobacco from Necta's garden. Tobacco smoke to the Shuara was as important as a good machete. During ritual sessions the men took turns blowing tobacco smoke through bamboo tubes into the other's lungs. A Shuara man is required to swallow the smoke he receives in order to achieve greatness as a hunter. Green tobacco water was used in many rituals for the curers and the bewitchers. On an expedition into no man's land it becomes an invaluable tool for masking the odor breakfast, lunch or dinner, especially if one has eaten lamb, beef or pork. It also deterred certain insects from swarming and biting, especially the black bee whose bite was agonizing and caused painful knots to form on the skin.

They had enjoyed their rest break and strong tobacco when without warning they were jolted back into the moment by the distant blast of a 12 gauge that stilled the naturally noisy jungle. They immediately reacted with concern; they knew that Wajai was signaling them to come quickly. Had they found the rock open in the middle? Were they in danger? Without hesitation they began the perilous journey back across the river they had just forded.

Some distance away Ikiam had slowly lowered Wajai with the rope vine until his lower torso had reached the mouth of the cave. Suddenly out of the cave lept a young 200-pound jaguar. The cat slammed a powerful paw into the right leg of Wajai, causing the startled Shuara to begin spinning like a top. He frantically held onto the twisting and turning vine. The surprising jolt had ripped the vine from Ikiam's strong grip and was tumbling down the incline, desperately grasping for anything to break his fall. The only support available was the pushi vine holding his friend. He grabbed it with both hands and hung on for dear life. It stopped his decent and the spinning motion.

Wajai bobbled like a cork on water while Ikiam dangled a few feet above him. Wajai struggled with one hand to free his shouldered shotgun. The jaguar turned toward the men, snarling and preparing for another charge. Suddenly the young jaguar slumped in his tracks and lay down on all fours cocking his head, examining the strange intruders. The cat's curiosity kept him occupied as Ikiam lowered himself enough to wrap his legs around the waist of Wajai, giving him the needed support while he tried with one hand to slide his shotgun into place for firing. Ikiam's quick thinking managed to limit the wobbling effect. But they were rocking back and forth which placed more strain on the brittle vine. Wajai managed to cock the shotgun and fired for effect rather than the kill. The shot startled the curious cat, which beat a hasty retreat just as the pushi vine snapped and the two men fell in a heap at the base of the cave. Wajai lay unconscious and Ikiam's right knee throbbed with pain. When Wajai came to, both men started laughing hysterically.

Necta and Yaupi were moving as quickly as they could. They cut a trail up to the mouth of the cave and found their two friends slumped in pain and animated conversation. Ikiam massaged his knee and Wajai rubbed the egg-sized knot on the side of his head and, after a lengthy discussion peppered with laughter, they all decided that exploration of the cave could wait for another day.

After retrieving Wajai and Ikiam it was time to get the injured men back to camp and prepare for the slow journey back to Sucua. Yaupi cut a tree branch and fashioned a makeshift crutch for his injured friend while Necta drew a crude map of the cave's location by studying the terrain and location of the afternoon sun. The men marched single file back towards camp. Necta silently wondered why the cat broke off its attack? It was very strange, but everything connected to this expedition was unusual.

The wait for a weather window to fly was again taking its toll. Bill spent his spare time sharpening knives, oiling weapons and preparing for war. "When the bell rings, you better have your shit together," he reminded us. We were due to join the Shuara in a few days and all of us were raring to go. Our thoughts were constantly with Necta and his team. Each time Bill thought about our guys out in tiger country armed only with one AR-7 rifle and muzzle-loading shotguns, he rolled his eyes upward. "Hoss, these guys have got to be the bravest men, I've ever heard of. Shit! If the Marines had any sense they'd recruit a couple hundred of these warriors, send them to Nam and we'd control the jungle supply routes in a matter of weeks." Bill was right, providing the Marines could control the Jivaro urge for taking tsantsa, enemy heads, shrinking them, and capturing the souls for their own use.

Jim had brought down a certified cashier check to fund our expedition up the Tutanangosa and give us staying power. Unfortunately, the banks in Cuenca were not impressed with an American cashiers check and insisted it be returned to the bank drawn on for proper identification. The only option remaining was to roll the dice with our meager funds. We agreed. Bill would haul the dredge to the Zamora, a gold bearing river in the Oriente. Alfredo, who claimed to have worked a small gold-bearing tributary upstream from the Zamora, assured us it offered the greatest promise to replenish our depleted funds. The plan was in place: At the Zamora they would rent two canoes to take them upstream to work for seven to ten days. Everything was scheduled and ready to go.

Everyone seemed upbeat when we finally arrived at the airport for Bill and crew to board the DC-3 flight that would take them to Gualaquiza, a small village that contained an airstrip. The flight saved days of overland travel and once there they could rent horses for the seven-hour trip up to the Rio Zamora. The manifest consisted of twenty-five pieces of equipment, including dredge parts, sluish box, and dredge, cans of gasoline, water, food supplies and personal items. Adriano and I wished them good luck and headed back to the hotel so that we could prepare for our flight to Sucua.

It was high noon when we crawled into the charter craft, and strapped ourselves in. The teeth-rattling, mind numbing take off was always the same when the roar of the engine mixed in with the sound of small stones peppering the craft's skin as it stormed down the runway. We circled over the wide valley of Azuay province and gained the necessary elevation to clear the high mountain ridges as the small red-tiled roofs of the farmhouses shrunk in the distance below. To the south the many green fields below merged into one giant green mass and the towering Andes temporary hid in the puffy clouds that locked in our serenity.

Suddenly a cargo plane came into view only a few hundred meters in front of us. Everyone yelled at the same time including our pilot who banked hard left with lightening fast reflexes that managed to escape a head on collision.

"Damn," I yelled to no one in particular as beads of cold sweat popped out on my forehead. I looked over at Adriano who had his head in his hands and then I glanced down at my left hand, which was twitching to an uncontrollable inner beat. "Damn, how many times can we beat the odds," I thought.

Everyone in the small Cessna had ashen faces and shaky nerves when we finally landed on the bumpy, grassy runway in Sucua. We couldn't wait to get on the ground and when we did, it was a while before our stomachs joined the rest of our vital parts.

We walked silently down the dirt road toward the Jivaro settlement. We passed a row of papaya and banana trees to our left and steered clear of a man on a mule dragging some heavy logs for construction of his house. The soldiers stationed at the makeshift airport control center, eyed us suspiciously for a brief period of time but soon lost interest and continued their card game. It was hot and we were sweating profusely, but it was a welcome change from the chilly weather in Cuenca and gave Adriano some respite from his nagging cold and runny nose.

By the time we reached Necta's house some color had returned to our faces and our conversation was focused on the task at hand. We brought a carton of American cigarettes, batteries for the portable radios, candy for the children, and bad news. We informed Esus that due to uncontrollable circumstances we would have to postpone the pending expedition up the Tutanangosa. We were relieved when he didn't bother to ask what the circumstances entailed, perhaps he could see the strain of the flight written on our faces and didn't dare ask.

Necta's wives seemed mesmerized by something in our demeanor and were giggling with each other when their husband arrived. Slung over Necta's shoulder were two howler monkey skins that he brought as gifts. In one hand he held two iron wood lances and in the other, a monkey skin purse complete with shoulder strap. It was the Jivaro way. If someone brought them gifts they in return would give gifts. When I presented him with two new backpacks, he presented me with a bodecaro, (blowgun) and a gourd filled with poisonous curare and darts.

We were spellbound by his amazing story about their adventures with whistling snakes and the pack of predatory tigres. He explained that more time for exploration would be required because of the thick trail-less jungle in the area around the Santa Rosa. One day of searching was limited to hundreds of meters not miles. Daylight hours were quickly used up trying to establish trails along the riverbank and nearby mountain ridge. The jungle was so thick that it obscured any landmarks unless you just happened to stumble on them by accident and there were too many derumbe (landslides) to count. In his opinion the landslide clues left by Mejia were useless in today's world.

When I gave him the bad news that we had to postpone our expedition and return to Cuenca, he became noticeably silent. Oddly, Necta seemed relieved we were postponing our trip. Having heard of his ordeal in the Macas forest and the injuries to some of his men, I understood his relief. Had I understood that he believed the predator attacks were the work of a wawek and his uwisin power was in question, would I have had second thoughts about continuing our search? That question was unanswerable.

When I returned to Cuenca, immigration agents were waiting. They informed me that I was in their country illegally because my visa had expired. I was ordered to leave Ecuador within three days or be placed under arrest for violation of Ecuadorian Immigration Law. My mind went numb. What would happen to Bill and my crew in the Oriente? They hadn't a clue I was being kicked out of the country. What would they do for funds if they don't find gold? What would Necta do? What a mess!

CHAPTER 10

DESPERATE MEASURES

The Rio Zamora

A full moon cast strange shadows across the dugout canoes along the beach of the Zamora River. There was an uneasy, eerie feeling among the men. Bill labored with his ragtag crew as they struggled to position their heavy loads of equipment into two dugout canoes poised to take them down river. Items passed along the human chain totaled twentyfive pieces, including packs, gasoline cans, the dredge, food and their personal weapons.

Maintaining an acrobat's balance on the bow of the dugout that bobbled up and down in the black shapeless river, Bill hoped that no mishaps would stall their progress. He checked the weapons, then his watch, then the murky water, knowing full well that darkness on the river could spell danger if there were any large tree trunks floating in their path. One bad mishap with the boats could lead to a loss of equipment and supplies and perhaps loss of life in the fast flowing muddy water. Oriolfo and Gonzallo wrestled to secure the six-inch alluvial dredge while Alfredo studied his map.

Pure Semper Fidelis, Bill Wright was a former Marine Corps sergeant proficient in combat training. Lost in his own thoughts he peered into the muddy bleakness of the swirling current. He massaged the nasty scar on his right leg that often reminded him how lucky he was to be alive.

He had joined the U.S. Marines right out of high school. "It was the right thing to do. There was a war going on," he said to anyone curious enough to ask.

Before he was dry behind the ears he found himself leading a squad of grunts on a reconnaissance patrol in the back woods of North Korea. It was there that the young man from Fort Payne, Alabama, graduated to manhood when a bright burst of light gave away their position on a lonely ridge. A mortar round zeroed in on the grunts and separated he and his buddies. He only remembered thinking he was too young to die and then passed out from shock and loss of blood.

When he regained consciousness he was riding on the shoulder of his old gunny sergeant who braved enemy fire to rescue the fallen Marine. The regiment had been cut off and surrounded by Chinese and Korean forces in a conflict that later became known as the "Forgotten War".

Dragging on a cigarette, he choked with a twinge of emotion thinking about those perilous days and nights and all his fallen comrades on the distant shores of Korea. He smiled to himself when he remembered the Marine Commandant who had been given orders to retreat and uttered those historical words, "Retreat hell! We're attacking in a different direction."

He remembered his old Gunny sergeant and wished that he were here with him. "He would have enjoyed the adventure", Bill thought.

Bill learned to box in the Corps and had won lightweight and welterweight fights, which resulted in a string of trophies and respect from officers and enlisted men alike. The Corps had given the red neck from Alabama a sense of selfrespect and a home that was unlike anything from his childhood. He missed the Semper Fi that the Marines instilled in the men that called the service home.

He said he couldn't remember why he left the Corps, only the fact that Sergeant Black called him pogey bait for not reupping and warned that civilian life would be his downfall because he was a warrior at heart. But, civilian life offered greater financial opportunities and the potential for personal success. Immersed in the real world he quickly learned that with only a high school diploma his opportunities were limited. He worked as a day laborer spreading concrete and cement for road builders, drove water trucks, oil trucks, and eventually went to the bigger rigs before finding the right girl to settle down with and have a family. And, like the men in the dugout with him, he, too, was searching for his personal El Dorado.

Bill eyed Alfredo in the pale moonlight. Hope and suspicion mingled in his mind as he watched the Spaniard twist and turn his map. The expedition came together because of this man, who claimed to have retrieved gold from a tributary that flowed into the Zamora and who now seemed to have difficulty reading a map.

The Rio Zamora was an ugly, big brown river that flowed from west to east along side the rugged Condor Mountains, which formed a natural barrier from the Huambisa territory that was the home to the Jivaro of the same name. The Huambisa still practiced the taking and shrinking of heads whenever the spirit moved them. Alfredo warned the crew about them, as the weapons were stacked on board the dugout.

Gold was plentiful in this remote region of Ecuador and with the six-inch dredge to replace sweating men and countless hours of backbreaking work, Bill felt their prospects of returning to civilization with a large amount of the precious metal were excellent. They would be, after all, on the fringe area of the legendary seven cities of gold that the Spaniards had searched for centuries ago. Legend was that the cities were made from gold; it was a story that led many naïve explorers to their early graves. Bill and Alfredo were both smart enough to know that the legend of El Dorado was distorted, yet they believed that the cities had existed in some form and produced huge amounts of ore at one time.

The dugouts strained under the heavy load of men and supplies as the moorings were dropped and the canoes began to drift in the turbulent water. The outboard motors roared to life stabilizing the wooden boats on the roiling brown river. Alfredo told me later he breathed a sigh of relief as they picked up speed in the dark night and headed downstream to the fourth small tributary on the right. He was happy his telegram to rent the two dugouts had arrived in time otherwise their expedition would have been doomed from the start and Bill would have had his ass.

Alfredo had met many Americans in his life but Wild Bill Wright was different from all of them. The man that sat on the stern with the brimmed hat pinned back with a Marine Corps emblem was something else. With Bill at the helm Alfredo said he felt like he was going to war rather than mining for gold. Instead of headhunters lurking in the jungle he half expected any moment to be in a firefight with the Vietcong. It was unnerving, yet somewhat comforting, to know that it would be virtually impossible for anyone to sneak up on them with Bill in their midst.

The constant hum of the outboard engines on the two oversized canoes making there way into the night gave the men on board a chance to relax. Alfredo could barely make out the ridgeline of the Condor Mountains that loomed before them, but he knew that beyond lay the three legendary rivers that defied previous attempts of exploration.

The Upper Maranon, the Santiago, and the Morona were well known to the Indians who lived in the area and a few brave Jesuit priests who went there to save souls, but few others had tested their machismo by scouting these areas. Those that had survived, at least according to rumor, came out wealthy. Alfredo shared a story of one Juan Salinas, a Jesuit friar, who traveled the Santiago River and the Pongo River in 1557. He had found gold.

Alfredo had learned through church records that the quantity of gold brought out from neighboring areas were so overwhelming that Spain took every measure to hide its operations. He learned that from one placer site worked during the years 1558 to 1652, the Spanish Crown received 400 to 600 kilos of 23karat gold. On another site, records showed that in each of the four consecutive years from 1570 to 1574 the Crown recovered an average of 9,500 kilos, about 25,000 pounds. What was amazing was that the Crown's legal share was only one-fourth of the total take. The mines had been operated secretly and no one knew what the actual total recovery was, only what was reported.

He believed that the legend of El Dorado was founded on real mining enterprises and not fable. In his mind the Seven Cities of gold that had been unsuccessfully searched for by Coronado, Vaca de Cabezza and Sir Walter Raleigh were in fact seven actual cities along the Amazon headwaters that the Jesuit order and the Crown protected from outsiders.

It was commonly thought that after the seven cities were established, the church's secret agents fomented wild stories about cities made from gold. They gave misleading instructions so that freelance explorers and conquistadores found themselves on a perpetual wild goose chase in hostile territories. So successful was the church deception that the only Englishman who ever tried to find El Dorado was Raleigh and he went up the Orinoco River instead of the Amazon. If these theories and rumors were correct then this was unquestionably one of the most successful pieces of counterintelligence deception ever accomplished in the history of international intrigue. But then, gold seemed to bring out the best or the worst in everyone.

Alfredo never really trusted the Church although he converted to Catholicism. He never shared his reasons for converting but in his core he was a true conspiracy theorist who believed that the Jivaro, Necta, had found the emeralds. On the off chance Necta had not found them, Alfredo was certain the gems were near the Salesian mission that had bought up a large tract of land and subsequently blocked attempts to explore the south bank of the Tutanangosa. If the Jivaro had found them then no one would find the emeralds because Necta would take everyone on faulty expeditions. If it were the Church that had made the discovery, then those who trespassed would probably experience unusual harassment and assorted problems in their lives, all brought about by the powerful religious group. His money was on Antonio Necta because nothing unusual had happened to he or his family.

Alfredo had never found his El Dorado, which, in his mind, forced his wife to toil long hours at the Cuenca telegraph and phone office to support his addiction to exploration. He lacked self worth and unwittingly sabotaged anything he tried to achieve. He dreamed of moving his family to a finca, a ranch, on the outskirts of Cuenca to escape to a more quiet life in the serenity of the mountains. But it was only a dream and somehow that made it safe. This trip was different; he was close, so close to fulfilling his dream of returning home a rich man. His wife could quit her job and he would gain confidence and respect. The dream was within his grasp and being navigated by a U.S. Marine.

On Alfredo's instructions the dugouts swung right and entered calm water leading to a sand playa that would be their home for the next ten days. Bill was the first to jump into the cool water. Grabbing the towline he noticed that low hanging clouds were blocking light from the moon. He dared the weather to interfere and barked orders for everyone to make haste. The men hustled to get the equipment ashore and set up the tents so their tired bodies could rest. Once the dugouts were unloaded their pilots would take them upriver to their births and return in one week with new supplies. The men waved farewell to their lifeline and settled in.

Before long the tents were secured and the men were paired off in the three tents along with weapons, food and supplies. Bill took one tent while Alfredo and his nemesis Oriolfo were assigned another. Gonzallo and two others tossed their bedrolls into the third tent and everyone opted to forgo food for the sake of rest. Bill checked his watch. It was 10: 05 P.M. They had left the Hotel

Cuenca a little after 8: 00 A.M. and had been on the go for 14 straight hours. They were spent.

Camped on the edge of the Amazon jungle the men quickly discovered that they were uninvited guests in the place. It was as though every indigenous living critter had ceased their normal routines and waited until the explorers were almost asleep before renewing their timeless patterns of pesky behavior and obnoxious communication. An incredible clicking cacophony that sounded like thousands of Spanish castanets came from the cicadas that competed with the drum booming of the huge-throated tree frogs. Soon the chorro monkeys began to hurl down their displeasure of having strangers on their turf.

About the time the noise became bearable in came the dangerous mosquitoes, some as loud as a vibrator. Immediately tent flaps were slapped down eliminating free passage to fresh blood, but also blocking out the cool evening air. The choice was easy: sweat in stale air or battle the insects. Both were miserable options for sleep. In time the tired men fell asleep only to awake at sunrise sweaty and clammy, but somewhat rested and ready to coax the tears of the sun from the small tributary that was brownish-red in color.

Gonzallo prepared breakfast while Oriolfo and another Indian cut firewood. Bill began the tiring ordeal of pumping up the large truck inner tubes that buoyed the mechanical beast that he had already assembled. Alfredo wandered upstream with batea in hand and began to sample the river.

By noon the Briggs and Stratton engine roared to life breaking the stillness in the lagoon causing the howler monkeys to come swinging through the trees to see what madness these strangers had brought into their world. They curiously watched the men wade chest deep in the water of the slow moving current, and guide the six-inch suction hose around the small bend in the tributary. Alfredo led the way after bringing up color near the eddy.

"Aqui, Aqui!" He screeched.

Mooring the dredge where Alfredo pointed, Bill grabbed the thick hose and began a sweeping motion on the rocky bottom of the river. "Just like vacuuming my living room floor," he yelled.

Alfredo's eyes were wide with wonder when he saw the sluice box fill with water, dirt and small stones. But the men's joy was short-lived when it became apparent the six-foot sluice box was no match for the massive amount of river residue cascading over the riffles designed to impede the flow.

"Stop the engines," Bill yelled. Everything came to a screeching halt. The dredge listed to the right from the uneven weight of its content and Oriolfo scrambled to keep the tubes from twisting out of their harnesses.

"Damn," Bill mumbled. "There's too much water and not enough sand."

Bill didn't use any curse words except one, "Damn", and that was used sparingly and only as an extreme reaction. On that day he set a record with the word. It took all afternoon to clean the dredge and then to experiment with different throttle settings and angles to regulate different air pressure in the tubes floating it. It was a long day.

By late afternoon frustration replaced the early morning excitement as the men continued to test the dredge's settings with no success. The sluice box needed an eight-foot extension to slow down the water and any sand containing gold that kept barreling through. The closest sluicebox extension was seven hours distant by horseback and forty-five minutes by DC-3, providing a tinman could be found to make one. There were other problems, too.

The river bottom was littered with large rocks that required a man to work halfsubmerged in the cold water while guiding the suction hose around the obstacles. Also, the bed of the river was a mixture of pebbles and sparse sand, which meant there was an uneven flow of small rocks and river water through the sluice box. By the end of the day they only had a few grams of yellow to show for their backbreaking work.

Tired and disappointed the men sat on the playa next to a roaring fire silently picking at their food. The dredge anchored in the shallow water cast an eerie crocodile-shaped shadow when illuminated by the moon. Alfredo sat alone, silently cursing the day he was born. He felt as if his whole life had been jinxed and questioned his continued search for that pot of gold.

Suddenly, without warning, straight off a five-foot bank sprang a muscular jaguar that apparently had been attracted to the camp by the sweet, strong smell of pork cooking in steel pots. The massive yellowish and blackrosette colored beast struck the water and quickly surfaced with powerful leaping motions that made it appear it was walking on water. Bill said he ran to his tent to fetch the 44-magnum carbine and collided with Oriolfo who had heard the commotion outside and was in the tent trying to put his pants on. He had one leg in and the other out, doing a jig, when Bill bounced off him, spinning him sideways. Alfredo tried to find his AR7 while Gonzallo and Oriolfo ran to find their machetes.

The jaguar turned toward the dredge and curiously eyed the black silhouette. Bill opted for the double ought load in the sawed off shotgun rather than the 44 magnum. He began yelling for someone, anyone, to hit the cat with light. Soon the river was flooded with flashlight power and the cat headed towards the far bank just as Wright cut loose.

"Damn that hurt!" Bill yelled in pain caused by the powerful recoil of the shotgun as steel pellets cut a swathe through jungle foliage and ricocheted off rocks on the far bank.

It was the first laugh they had all day. They huddled around the campfire to finish dinner and coffee and cradled their weapons and wondered if the cat was still running?

The following day the problems continued with the stubborn dredge. It refused to regulate the flow of water because of limited sand. It had terrific suction and was working between eight and ten yards of material an hour, but the sluice box was too short to handle it. The dredge problem was compounded by black bees the size of a man's thumbnail that were attracted to the sweaty men.

The bites were painful and left a large knot that itched and then festered. The repellents brought with them didn't faze the nasty insects that swarmed in dark funnels several feet in the air before selecting their target for the day.

Fourteen grueling hours were spent wrestling with equipment, swatting bees and trying to work the cold river's chill out of their bones. Bill wished he had allowed some whisky or native brew to be brought to quell the chill. He refused after being cautioned that Alfredo's weakness was booze and that it fueled his demons. He hadn't wanted to take any chances and opted to keep the camp dry. He told me later that he regretted his decision and that a long swig would have eased a lot of woes.

They had been on the river for two days with nothing to show for their work and were ready to cut their losses and go home. Only one problem; it was impossible due to their logistic nightmare. The dugouts would not return for five more days to take them back upstream to horses and mules that would carry them back to Gualaquiza. They quickly realized they were stuck on a non-productive river with nothing to do accept wait, something the Marine didn't do well. Bill decided they would try to find placer outcroppings using picks, shovels and a few bateas to try to salvage the operation.

The local Indians shared more depressing news. The small tributary that Alfredo guided them to whose mouth opened into the Zamora was designated a red river. Those rivers were also known as "Starvation Rivers." The waters were reddish brown, which most thought was the color of silt, however the Indians brought in from Gualaquiza to work with them explained that the river was dead because it had been naturally poisoned by a copper deposit somewhere higher up. There were no fish, which they had planned to add to the menu, but birds were plentiful so Bill supplemented their food supply with fresh bird meat. For the remaining days Bill hunted alone. Alfredo knew the Marine was angry at his decision to bring them to this place, and he lay low, giving Bill wide berth until the dugouts returned to pick them up.

Following my expulsion from Ecuador, I settled back in Phoenix, angered at fate for keeping me from the action. It was during this time that I met Mickey Moore an interesting man and a successful real estate salesman and entrepreneur. Mickey was also into the New Age beliefs, organic foods, supplements and psychics. He was an ex-bodybuilder with strong and well-defined muscles that complimented his large frame. His rosy cheeks made him look much younger than his true age, which was mid-forties. I was intrigued when he asked me to join he and his wife at a special place for dinner that featured a unique psychic act. I wasn't sure I believed in any of that but I happened to have the Mejia book with me and thought it was worth my time to see if this psychic could unravel the riddles the Colombian left as his legacy.

The Blacksmith Shop was a quaint restaurant in downtown Phoenix that reeked of mystery and intrigue. Constructed on the original burned out site of a turn of the century blacksmith shop and hotel, the resurrected eatery was rumored to house a resident poltergeist. The friendly ghost was well documented by numerous employees and had apparently taken up residence in the restaurant that featured good food and a psychic named Richard Ireland. Ireland was packing the place every night with his uncanny ability to utilize psychometry and ESP on stage.

Ireland's act began by placing gauze over his eyes, then wrapping tape around his head, and finally positioning a black hood over his face and eyes. He asked members of the audience to send up an article of clothing, or personal objects or to write questions on a piece of paper so that he could zero in on information they requested. I sent up the last will and testament of Raphael Mejia that was bound with a shoestring from Mickey's shoe as added protection against anyone's prying eyes. Ten minutes passed and Ireland accurately "read" scarves, earrings, rings and other assorted items to the delight of the crowd. We were almost through dinner when my name was called. I clicked record and taped the following.

"Hello Lee! Is this your shoestring that's tied around the book? (No) *Are you from Ecuador?* (Yes) *You are looking for something valued at*

$12,800,000...and I somehow feel that you are going to find pieces before you find the whole of it. It has to do with Victor, all the way down to like Teranzo or something like that. (Ireland was picking up the two rivers named Victor and Tunanza!) *Lee, would you understand that?* (Yes!)...*And this is scattered somehow, it's not all just in one place as you had thought. I get the feel of water...what you are looking for has to do with water...water by the gate or water by the door, water by the arch...water by something."* He paused for what seemed an eternity.

"I feel as though you are going to find it in connection with water... and in the proximity of a gate, a door or arch, or something of this kind. I don't know what that means, but I guess you would understand. I just feel as though you are not on a wild goose chase, so don't give it up...it's like you are heading in the right direction. I will have some one take the book back to you, if it is worth $12,800,000 dollars."

To say that I was impressed was an understatement so I arranged for a private reading the following day and once again I took my tape recorder to document the reading. Dr. Ireland began by closing his eyes. He seemed to enter a trance state as he held the old closed book in his hands.

"I close my eyes and see you moving one step at a time and I do feel that you are moving into the right area, breaking as it were the code... and you are beginning to find what you are looking for. I see a turban like a sultan's hat with a plume on it and I feel this may be...ah...some kind of a symbol...like the shape or the form of what you are looking for. As though it is like a turban with a plume on it...I did feel as if you would have...ah, seven days pops into my mind. I do feel that you will be able to slowly unravel this. It's not something that you will be able to do objectively from here. You must go back and unravel it from there. It's like a fog...you can only see one step at a time.

"I feel that you will find this as 'relating to'. And I feel your relative is not like a person relative but instead is in the idea of relating or relationship. As though this is relating to that and I feel as though you are going to get this and this will help you relate to that...as though you need to decipher and discern the writings and the things which

apparently you are uncovering. That will help you to uncover this. It will continue to unravel as you go along.

"I do feel that you need to be cautious, as there are other forces and powers against you. I'm speaking of physical powers and spirit forces against you. I want you to be careful! Move with wisdom and caution. Bodily danger involved, but you can circumvent that by using good wisdom. I want you to keep daily accounts of what you are doing. For some reason maybe this will be the foundation for a book you will be writing or maybe a film. That would be for you a lucrative thing and offer you fuller maturity. Please pursue spiritual avenues.

"Don't overlook the gold as I see it lying in the water. The river is not deep and is clear. Make sure you have all the necessary permits and credentials and I see no problems for you in bringing it out and selling it. There is no tunnel. I see a landslide, yes a cutaway and at the bottom there's a large rock that looks like a sultan's hat. The more I see, the more I believe what you are looking for is going to be there.

"Mejia...is the name that comes to me and I see him as a very orderly person who was not deceptive. He was very honest...very precise... very exact. He never did anything halfway and always planned ahead when called upon to do so. I feel sadness in the book. There was a lack of accomplishment. Like Moses coming to the Promised Land and not being able to enter it! The native people themselves are going to be your greatest help. I see an old man...a very old man that could be of help to you. Go with God!"

This was my first face-to-face meeting with a psychic and I was blown away by Ireland's accuracy. He not only named the key rivers, Tunanza and Victor, but their location to each other. Between these rivers on the opposite bank flowed the Cunguinza or what Alfredo adamantly called the Santa Rosa. Was Ireland confirming Alfredo's hunch? Ireland saw the landslide or as he put it, a cutaway and at the bottom a large rock that looked like a sultan's hat.

Was it possible that I now had an accurate description of the rock open in the middle hiding the tres arrobas of valuable stone? When Ireland mentioned "relating to" had he deciphered the riddle where Mejia said, "Like making a relation the entrance is by the..." This made sense to me because the riddle, "like making a relation..." depended on one clue relating to another in a sequence of discoveries.

Ireland did not see a tunnel or cave in his channeling. This might rule out our theory that "the entrance is by the cascabel" was a tunnel or cave. Was the gate, door, or arch he saw the cascabel? If so, then was this a natural landmark that signaled the searcher they had entered the location of the rock open in the middle that Ireland described as in the shape or form of a sultan's hat? He had seen much more adding new mystery to my quest. Each piece of the riddle that was brought to light left new pieces awaiting discovery. One of those pieces bothered me.

Ireland had issued an ominous warning of opposing physical and spiritual forces. He had cautioned me to go with God. I didn't have a clue what he was referring to. At this juncture in my life ignorance was indeed bliss. But I was curious about the identity of the old native man that might be of help to me.

Bill Wright was ensconced back in the refinement of Cuenca. He was happy to see me and excited to learn that our funds had been replenished. He remarked that he would kiss a leper if he could think and speak in English. At the airport he'd learned that three American miners were missing in the Oriente. Bill thought they were dead, killed by either two-legged or four-legged locals.

Reinforcing what I already knew he blurted out, "Hoss, this is the Wild West down here!" He laughed and colorfully brought me up to date on his adventures on the Rio Zamora.

Bill and I were more than ready to head for Sucua and the next leg of the Mejia adventure. The sun was peaking through the scattered clouds as the six of us, Adriano, Bill, Oriolfo, Rodriquo, Alfredo and I, stood at the end of the runway trying to look as inconspicuous as possible to the military personnel milling around the airport monitoring anything unusual or suspicious. Our pending flight to Sucua couldn't have been any more noticeable. We'd been there for four hours with eighteen bags of cargo that weighed 597 pounds, 300 pounds over weight. We clustered together waiting for the DC-3 to arrive from Loja when a jeep with two soldiers passed by, made a loop and came towards us.

The Ecuadorian army sergeant eyed us cautiously and commented to his driver, "I wonder if the Gringos have a permit for the shotgun."

Bill was wearing his brimmed hat with the Marine Corps emblem with his sawed off shotgun slung over shoulder. He made eye contact with the curious soldier. Bill didn't have to say a word; his eyes did the talking and the Ecuadorian sergeant suddenly decided it was time for a coffee break and the jeep sped off. While the rest of us laughed, Adriano remained passive; he knew that the military and the civilian government had a fragile alliance. The military had stepped down from power three years earlier but could reclaim their throne anytime they desired.

The National Air Service DC-3 with a black nose and yellow stripes glided in, touched down, and slowly made its way down the tarmac. This was our cue to strap it on, suck it up, and get ready to board an old airplane that was flying before I was even born.

The first DC-3 rolled out of the Douglas Aircraft Company on December 17th 1935 and was affectionately named the Gooney Bird. Those old DC-3's were the workhorse of U.S. aviation and a fine tribute to the American engineers who designed them, but by now they were tired from flying countless sorties around the world. Even more unnerving, the mechanics that worked on them appeared to be in their teens. In the past two years ten aircraft had gone down and I wasn't even counting the small aircraft and helicopters that were reported missing. The only thing certain was uncertainty itself as we prepared to board another white-knuckle flight to Sucua.

We entered the plane single file and immediately had to crawl over cargo to get to our designated area that consisted of four rows of seats in the back of the plane. While we settled in Adriano made a comment that was ill timed for the moment.

"See the cargo loader over there," he pointed in the direction of a man, dragging equipment into the plane. "He was on the DC-3 that crashed in the Oriente and was the only survivor. He said it took him five days to find his way out. When he finally got back he was minus one pound of skin and three pairs of pants."

Adriano grimaced while Bill and I just stared wide-eyed. When the cargo loader bent over to arrange cargo his shirt rode up displaying ugly red welts covering the width of his back. He had survived the impact of the plane crashing through the jungle canopy, had almost died trying to escape the jungle without a machete or survival gear, and paid the price from the pounding his body took from the sticker vines. Everyone else on board had perished on impact.

He was one of those with that mystical white light around him, that unseen force protecting those whose time hadn't come. You know how it is? You want to be one of those chosen ones however you're never quite sure if you are. The young Ecuadorian shoveling cargo into the DC-3 was one of those.

Suddenly the DC-3 lurched to life. Engines revved and smoke belched as we hurled down the runway gaining speed and tolerating the unbearable noise from the door up front that was rattling like crazy. I remarked to Adriano that I hoped the door stayed closed until we landed in Sucua. It was my way of paying him back for reminding me that the Ecuadorian skies were indeed not friendly. We gained altitude and soon were over the backbone of the Andes, entering a milky-white sky with low visibility.

The Andes were formidable and ranked second to the Himalayas in ruggedness. I hoped I would never have to fly through the Himalayas because these breathtaking, majestic mountains were an utter horror to navigate over, around and through the narrow passes that led into the Upano valley. No matter how often you flew, you never quite got used to the air pockets that appeared out of nowhere and caused the plane to drop dramatically in a second of sheer terror. Not only would you lose your breath and your lunch, your testicles took up residence somewhere in the upper body.

The Ecuadorian pilots were leery of the sudden and violent down drafts that sucked the aircraft towards the jungle covered mountain peaks. Somewhere out there were dozens of planes and helicopter skeletons lying beneath the triple canopied jungle filled with rotting cargo and deceased occupants whose loved ones still grieved and hoped that someday they would be found for a decent Christian burial. It was brutal to think about these morbid scenarios while flying into Sucua but I couldn't ignore the thought of that fragile thread that separated each and every one of us from living or dying.

Window seats offered a spectacular view of the majestic Amazon jungle with its winding rivers marching timelessly through the deep green basin where they eventually emptied into the ocean. It was awesome and inspiring and gave pause for thought. After my first couple of flights I shied away from the window seat. When the milky, opaque mist engulfed the plane flying below the jungle-encrusted mountain peaks I felt helpless and realized that I was riding in a stream of uncertainty. It was unnerving to know that my fate was in the hands of a pilot who was experiencing what I was seeing, which was nothing more than white space.

The pilots that survived were grateful for the instruments that guided them through the narrow passes and over the steep ridges. The two words that pilots detested the most were 'pilot error'. They considered it an insensitive accusation heaped on them by the press. Pilot error was the catchall phrase for the media in its explanation of the number of plane crashes in the country. Who could argue with the reporters? None of the pilots who sat up front survived and the wrecked planes were rarely found.

Today, we were lucky. The proud old DC-3 glided in for a three-hop, skip and bump landing on the runway filled with water. It had been raining continuously for three days in Sucua and, according to the locals; it was the beginning of the rainy season, which was forty-five days ahead of schedule.

"It's because of the long hot summer," a man at the landing strip told us as we unloaded our cargo.

We skirted the eight hundred pounds of fresh cut beef and fifty pounds of flies that were stacked alongside the plane for the return flight to the kitchens of Cuenca. It was good to finally be back but we all knew we had serious problems if the rain continued.

There were between eight and ten rivers to cross before we reached our final destination and if it was raining in Sucua then it was raining higher up in the mountains. Those little tributaries that you could normally wade across in December would be rolling with white water. Oriolfo and Rodriquo stayed behind to arrange a truck to haul our eighteen duffel bags to Antonio's house as we made our way down the muddy road, getting drenched and saying little.

Esus greeted us with a smile and invited us into Antonio's house to wait for him to return from visiting a friend. Esus hadn't changed much from the last time I saw him except that the wet weather seemed to aggravate his even more noticeable limp. Bill was never one to shy away from asking forthright questions and he couldn't contain his curiosity about the Jivaro's limp. He had the balls to ask what we all wanted to know but were too polite to question.

"How did you hurt your leg?" Bill blurted out.

Esus responded with a grin and then gave us a glimpse into his private world that Adriano translated.

"When you find the woman that you want to marry, in Shuara custom you first must court her and find out if she is interested in you. I found my woman and courted her then I sent a close friend to be my intermediary. My friend talked to Antonio and Muriyima to tell them that my intentions were honorable. My friend told me that there was no opposition. Then I was invited to Antonio's house. I was allowed to sleep in the men's end of the house. Before dawn, I crept out of the house with a blowgun and went hunting. I was hoping to kill many monkeys and birds to prove my manhood.

"I was successful. I returned to Antonio's house with all my fresh killed game and Antonio was impressed by my marksmanship. Then I offered the fresh game to my future wife, to cook for her family. Now comes the time that she must make the decision if she wants to marry me. When she accepts the monkeys and birds for cooking and squats down beside me then she has agreed to marry me. She took the game and cooked it. She joined with me in eating the food and from that moment on we were married. That same night her family allowed us to sleep together in their house.

"I had a beautiful wife so it was my duty to build a house for the two of us to live together in happiness. I selected property next to the Tutanangosa and began my work. Before long it was ready for us to move into. It's on stilts off the ground to protect us from snakes. We had one door that could be locked to protect us from the niawas and my wife wove window shades from the brava cane for our two windows so we would have privacy.

"The first night in our new home we made love. It was hot and we rolled back the shades from the windows to enjoy the cool breeze coming from the river. We were in the moment of lovemaking that is most enjoyable when a tigrillo jumped through the window and landed on my right leg and of course I screamed because all I could feel was claws in my leg. She did not know about the tigrillo. She had her eyes closed. She thought that I was finishing my lovemaking and began screaming. I was screaming in pain while she was screaming with joy! She was begging me not to stop. She had found a man who enjoyed primitive lovemaking that included authentic sounding snarls that the cat makes. I was struggling to free my leg from the angry little tigre's jaws. Finally, she opened her eyes and from the light of the moon could see that I was fighting for my life. She began screaming louder and startled the tigrillo. He let loose of me and jumped through the window."

Esus took a deep breath before continuing; his gestures animated the words. The four of us were holding back laughter while trying to maintain a straight face and show concern. Alfredo was choking. Adriano was making funny little snorting sounds but trying to preserve his and Esus' dignity. Bill turned to look at me but I knew if we had eye contact that we both would have exploded in laughter. I looked away, choking back the tears. I placed my hand over my mouth to try to suppress what was becoming impossible to control. Apparently our laugh suppression worked because Esus continued on.

"Afterwards when we made love, it was never the same. I think she had been spoiled from that first night in our new home."

That was it! That was the statement that rocked the hut with uncontrolled laughter. Bill had been sitting spreadeagle on the floor and was now on his back banging his head on the floor, yelling "Mercy. Mercy," in between thuds.

Alfredo hardly ever smiled but was laughing in a controlled and polite manner with low frequency sounds. Adriano, whose laugh was always in the domain of a controlled snicker, was beet red, tears flooded down his face as he uttered high pitch sounds that sounded like a person choking on food. I was laughing so hard that I thought I was going to piss in my pants. Then I looked toward Esus to see if he was going for his blowgun or shotgun. Instead he had a wide grin on his face, content that we were happy.

Esus had provided the humor that we sorely needed but we were in trouble and knew it when Antonio arrived soaked and in a somber mood that magnified our concern. He said the rivers were flooding and recommended that we leave all our supplies with him and return in March. Any expedition in this weather would be too dangerous for everyone concerned. He questioned getting our supplies by horseback to Wakani.

"Come back next month," he repeated.

Bill had now been in Ecuador for sixty-eight days and his three-month visa was winding down. It was either now or never for him and he was agitated by the thought of giving up. He didn't see the problem. Adriano felt a sense of responsibility for everyone and a heated argument ensued. He scolded Bill for not examining the whole picture and asked how he was going to cross the rivers. He demanded Bill explain how he could hunt for something when he couldn't see fifty feet ahead in a rainstorm. Bill backed down and suggested we take a vote to determine who wanted to go and who wanted to stay. Adriano agreed.

Bill and Alfredo voted to go up the Tutanangosa. Antonio, Adriano and I voted against going. But we all agreed to sleep on it. Then we left Antonio's house and headed for the local hotel to find rooms for the night.

The Hotel Pension was not too bad once you got used to the smell and the thick spider webs that crisscrossed the corners of the ceiling. The small rooms contained two beds, one nightstand, a small mirror on one wall, and no bathroom. The outhouse, which served as the community bathroom, was about fifty yards from the hotel and required a scary walk through a thicket to reach it. The clerk warned us to be careful, as a man had been bitten in the outhouse by a fer-de-lance. I imagined what part of the body had been bitten and immediately developed an acute case of constipation.

The hotel sat on the main street that was nothing more than ruts and potholes filled with standing water and mud. We were on the second floor and spent most of the remaining evening on the balcony checking out the street activity. Across the street a man from Colombia sold young tigrillos for eleven-hundred sucres each. There was an electrical repair shop where a caged bird danced to the music from a radio strung in a tree. The clerk warned us that when the radio went off, the lights would go out. He gave us matches and candles that we carefully laid out on the nightstand. On the street corner next to the hotel was a hole-in-the-wall restaurant that was open until the power went off so we hustled downstairs to get some food before they closed.

After dinner we retreated to our rooms. Bill and I were bunking together so Adriano and Alfredo joined us to discuss the day's events and consider our dwindling options. We had just entered the room when the power went off. We grabbed the matches and a candle and laughed at the puny flame casting shadows. Our meeting was more like a séance. The draw of El Dorado clouded reason and logic. Alfredo and Bill remained adamant that they would take the expedition. We asked that they remain flexible, especially if it continued to rain throughout the night.

After Alfredo and Adriano left for their rooms, Bill and I got into a heated discussion over adventure versus desperation. I told him he wasn't thinking about his life or his family and certainly not about the project and the safety of those that would follow and lead him into the Macas Forest. He braced himself then reiterated that his visa was due to expire in twenty-two days and he would be forced to leave the country. He was unyielding.

Ignoring his stubbornness, I attacked. "Look Bill. You know the story about Little Hell and what happened out there. We almost died because the damned rain swelled the river. We were cut off. If luck hadn't been with us, all seven of us would still be up there, dead. You have to place some value on your life. What about your family? Shit. What the hell is your agenda? I know Alfredo's is greed. He's got nothing to loose even if it means sacrificing his life. You still have options. You have a family. You can go home. You can go back to work. What the hell is wrong with you?"

He stared at me blankly, void of any emotion. I might as well have been speaking a foreign language. Bill looked away silently signaling me that his mind was made up. Betty, my sister was right. He was one stubborn guy. My mind was made up, too. I wouldn't be going with him. My adage was simple logic. Discretion was the balance to valor when one desires to explore another day.

The next morning Antonio and Esus joined us for breakfast. Antonio shook his head when I told him that Bill, Alfredo, Oriolfo, and Rodrigo were going to stay and make the expedition. I was hoping he would refuse to guide them. But Antonio said that he would honor his commitment and take them up the Tutanangosa. There was something about the way he agreed and the strange, concerned look on his face that bothered me more than the rain. It was obvious that he thought the expedition was illtimed, supported by bravado rather than logic. I knew he would have preferred to wait this one out, but he felt obligated to honor his commitment. At least the guys had someone trustworthy to lead them up river.

After breakfast I maneuvered Adriano and Antonio into a private conversation and assured Antonio he would not be held responsible for any mishaps. I explained that Bill (Guillermo), my brother-in-law and ex-Marine, was fully capable of taking care of himself. Instead of relaxing Antonio appeared even more concerned, which bothered me.

Bill barked that it was time to burn some daylight. We shook hands and hugged each other and said "Vaya con Dios". Go with God.

Despite my doubts, I wished them the best when they left the hotel with Antonio and headed for the staging area. Adriano and I questioned whether we should fly because it had rained heavy all night and a light rain was falling at the airport. The Sucua airport consisted of three shacks that sat side by side. The first was the SAN office, the second a police and military compound and the third a small hole in the wall kiosk that served soft drinks. There were no other passengers because anyone with any sense had opted to wait out the weather.

We'd been there for several hours before the National Air Service employee told us the plane was en route and we should present our tickets. A soldier walked up and asked where our friends were? Adriano explained that they had stayed to hunt some tigres. The corporal looked at us in disbelief and replied that we must be joking and waited for the punch line.

When he discovered we were dead serious, he snapped, "Then they must be crazy or fools to hunt in this weather." His comment punctuated the stubbornness of those bent on taking the expedition.

There is something about a heavy rain that can send shivers up and down your spine. We were both challenging nature in our separate ways. Bill and company were on the ground. Adriano and I were in the air. Airborne and circling for altitude we had a panoramic view of what was lying in wait for the expeditionary force. The big river rolled like an express train from the white water tributaries that fed into it. Big, ugly, black clouds hovered in the background threatening to release more rain. Our team didn't have any idea what they were up against. They were headed directly into the fury of the storm that was readying itself for round two.

My guilt over leaving the team gnawed at me. Even though I knew I had made the right decision I questioned it. The DC-3 was bucking wind and a light rain pecked at the windows, making it difficult to see. We continued our climb, slowly leaving the mist-covered jungle sprawl below us.

"Maybe they will turn back," I thought as I closed my eyes for a catnap, trying to avoid the smell from the large pile of fly covered beef stacked between the cockpit and us. Sleep was impossible as the plane lurched in the strong wind currents. The thought crossed my mind that the guys on the ground had better odds of surviving than we did.

The first indication of trouble came when Adriano tugged at my shoulder and pointed to a small village below us. "Lee, we have passed the same village three times. See? There, the one that has the small public square next to the river. That's the one. We keep circling around and keep coming back to the same place. This plane is in trouble."

I looked toward the open cockpit and saw the pilot standing, fidgeting with something above the copilot's head. I worked my way around the beef to the cockpit and asked the startled pilot if there was anything I could do to help? He curtly told me to return to my seat and buckle up. The look on his troubled face answered my questions. From the pilot's position I could see mountain peaks we were supposed to fly over. They were much higher than the trajectory of the DC-3 as we continued to circle, buying time and sluggishly trying to gain altitude.

The flight crew struggled with their problem while Adriano and I made a futile attempt to buckle up in seats that weren't even bolted down to the floor. We were literally flying by the seat of our pants in a meat locker masquerading as an airplane that was being flown by two pilots frozen in fear with the engines straining at full power. Standing in front of me, holding on to a hand strap was the cargo handler, the survivor of a previous crash and a solid reminder of the danger unfolding.

I took a deep breath and pulled my reel-to-reel tape recorder from my backpack. Adriano shook his head in disbelief. If I was going to die I wanted to record what I was thinking and feeling. I pressed record and began describing that we weren't climbing at all and questioned how long we could continue to circle before we ran out of gas or faced some other drama. I commented about the treetops below us that appeared to be getting closer to the belly of the plane. The DC-3 creaked and groaned with each pocket of air, which I described in detail as Adriano clinched the arms of his seat that was dancing across the floor.

The plane bounced and heaved from the strong air currents, like a ship on high seas. The cargo handler, the one with the light around him, made his way forward then returned from the cockpit and made a motion for us to hold on. He suddenly unlatched the door. The strong gust of fresh air was a welcomed relief from the rancid smell of beef, but the noise from the engines was unnerving. I tucked my tape recorder back into my backpack.

Then the pilot yelled for the cargo handler that I had named 'Light' to get ready to bolt the door. Then he banked the plane to its right.

"Oh shit," I muttered. A group effort was required if we were to get the door closed before a strong gust of wind ripped it off the airframe. Neither Adriano nor I were fond of heights and here we were forming a human chain and slowly approaching an open door that led to eternity. Light made an attempt to grab the door and secure it.

I was behind him holding the belt of his pants and Adriano was holding me likewise with one arm while the other arm was looped through a cargo strap from the ceiling. The DC-3 banked so violently that the door slammed shut while we tumbled into one another against the wall. The door was closed and not latched. Once again we made our way towards the unpredictable door and while Light struggled with the latch we fought the turbulence of the plane to stay upright. After a few minutes it was apparent the door was damaged and would not lock securely. The pilot ordered us to prepare for an emergency landing. I retrieved my recorder and slung it over my shoulder, and pressed the record button.

We grasped the cargo straps so hard that our knuckles turned white as we made our final decent into the Cuenca airport. Then without warning, the pilot ordered us to cover our faces! The landing gear was stuck and we would be bellying in. I asked my recorder how the old gooney bird with age-weakened rivets could stay intact on impact. As we approached the runway my concern subsided when I heard a loud clunk. The landing gear fell into place, but had it locked?

We glided in as trees whipped by underneath. I described the emergency vehicles racing along the runway. If nothing else, they gave me a sense of moral support. Then we hit the runway with such force it almost ripped our hands from the cargo straps. We bounced and bounced again. The landing gear held up through the impact. "Hail Mary" could be heard echoing in Spanish throughout the plane as the craft skidded to a halt. We were safe. I clicked off the recorder and took a deep breath.

We staggered off the DC-3. Light was scared and confused and had the panicked look of a man who had taken his last flight. Perhaps his torch that contained the mystical light had been relayed to me? At least I hoped so.

Before leaving the tarmac I turned and took one last look at the aircraft that was still the pride and joy of aviation crews worldwide. This old bucket of bolts had come through time and time again. I remembered that when I was six years old my grandparents had bought me a seven-inch metal replica of the aviation workhorse. It was my favorite toy. Talk about destiny.

CHAPTER 11

SEMPER FI

Bill Wright

A light rain fell from the black clouds as Bill and his crew saddled up and hit the trail in single file. Suddenly, they heard the drone of our DC-3 and paused for a smoke break and one last look at the plane circling above them. With the sound of the aircraft engines growing fainter, Bill checked on the five Shuara hired to carry cargo. They had been selected because of their muscular stature and reliable past performance, but most importantly, they could be trusted. The men seemed to have seventeen-inch necks and a work ethic that bordered on robotic. With the aid of a strap around their forehead they could balance over a hundred pounds of dead weight and haul it for hours without any complaints or straggling.

It was time to move on. The cargo carriers slipped into the head harnesses, while bags of beans, rice, pork, potatoes and assorted fruits and vegetables were carefully weighed and placed in each man's designated load. The tents and the tent poles were another issue. One man could carry a tent providing it didn't get wet, which made it dead weight. To eliminate this problem Bill encased the tents in plastic sheets. Alfredo and Oriolfo helped Antonio adjust the load on the three horses brought in for the first day's journey. Antonio and his daughter, Maria, rode one horse, Bill had a horse and Alfredo shared his ride with supplies.

Antonio and Maria lead the way as they plowed into the jungle trail that was thick with mud. Men and horses plunged into the gooey slime that was like quicksand as it began to rain harder. At first the rain was tolerable but it grew heavier, soaking the men and beasts, and anything uncovered. Soon the trail was running with water and getting slippery. Antonio's horse fell and threw he and Maria into the muck. Fortunately they weren't hurt badly. Antonio twisted his leg, his daughter was fine and the horse was spooked, but unhurt. The horses could not find solid footing so Bill and Alfredo dismounted fearing a bad fall might break one of animal's legs. They opted to pull the beasts of burden through the running trough of water that was their trail.

They finally reached Wakani before nightfall just as the rain came down in thick blinding sheets. Drenched to the bone they unpacked the horses and helped the cargo carriers out of their heavy loads that were then stacked inside the ranch house. Antonio and Esus carried buckets of water up from the river to wash the mud from the horses before bedding them down under a thatched roof building next to the ranch house. In the Oriente a horse was the most valuable commodity a man owned and it always received preferable treatment over those that rode them. This day was no exception.

Bill started a roaring fire and the men huddled around its welcomed warmth. Drying out they realized that they would have to wash off the thick encrusted mud that clung to their clothes. Just the thought of going back into the thunderstorm and making their way down to the slick riverbanks to wash themselves in the icy water of the Tutanangosa made them cringe. It had to be done before the mud hardened and so, one by one they left the warmth of the ranch house to clean themselves. They returned wet and cold, but clean. The large pile of dry wood fueled the fire during the stormy night as the men slept huddled together like a pride of worn out lions.

The next morning there was no sunrise. There was no sun. There was only more rain. Esus and Maria prepared the horses for the return trip to Sucua as Bill stood in the doorway of the ranch house and gazed into the murky fog and rain. He lit a cigarette and took a deep drag. He was questioning whether to continue or turn back. Alfredo joined him and both men relieved their anxieties through the calming effect of nicotine. Pushed by Alfredo's insistence, Bill made a fateful decision to continue on.

"Esmeralda's aqui," Alfredo whispered while drawing a crude map of the area in the mud as rain cascaded down his face.

Bill looked at him and then the drawing and remembered my conversation about this Spaniard who had nothing to lose and everything to gain. "What the hell," he thought. "Every man has a price."

Visions of green fire extinguished my warning from his mind. He stood and without any remorse in his voice barked, "Vamos!" to the waiting crew.

The phrase stirred activity and the men harnessed up their backpacks as cold rivulets of rainwater saturated the bulky loads making them heavier. The horses were turned back down the sloppy trail towards Sucua and the expeditionary force wheeled west toward a wet green monster lying in wait.

A black thundercloud rolled down from the headwaters of the Tutanangosa. It seemed to hover over the slow moving men and then, like obeying an invisible command from the rain god, it started to rain in torrents. The sound of the raging water from the Tutanangosa combined with noise from the heavy downpour made it impossible for the men to hear each other.

Communication was impossible. The trail began with a steep climb into the mountains and footing was precarious and dangerous. The higher the men climbed the denser the jungle became and before long the non-Shuara were struggling to catch up. Bill knew that any stragglers might mean a nice meal for a hungry predator. In a push of adrenaline he managed to catch up to the cargo carriers. He could only shake his head at the stamina of the Shuara that packed one hundred pound loads and all he had was a carbine, two bandoleers of ammo strung around his neck and a backpack. Before long his thoughts faded into the vanishing trail of tangled undergrowth.

Marching in single file they climbed up then slid down hillsides sweating and puffing. Then the exhausting ordeal started all over again until one lost count of how many times they had climbed and slid. They had already crossed eight small rivers that were newly formed from the heavy runoff. The water was knee-deep, icy cold, and swirled with debris. In a strange way this was welcomed; the running water allowed the men to wash the thick mud off their clothes and boots.

That afternoon they reached the first big river to be forded. It was the Victor and it was rampaging whitewater. Bill glanced over at Oriolfo and Rodriquo; they were holding their own. Alfredo was struggling and in pain. Ugly knots formed on his legs, yet he seemed more determined than ever. Antonio was not even breathing hard but concern was etched on his face when the men stopped to light up.

"God, I'll never smoke another cigarette again," Bill said gasping for air.

It took the better part of an hour for the men to safely cross the waist deep river. Near exhaustion they decided to make camp for the night even though it was still early. Three Jivaro used machetes to cut out an area for the tents as the rain continued to pour down. The welcomed relief from the trek through mud soon turned into dismay. Their beleaguered bodies begged for warmth and rest. Without dry leaves and wood there was no fire this evening, only rest. The other serious problem was their meal. Without a fire there was no warm food to help ease the chill from the wet and soggy clothes that clung to the cramped muscles of those carrying the heavy loads.

Drinking water was passed among the ragtag crew who munched on an evening meal that consisted of crackers and bread covered with canned sardines. Under waterproof tents the men huddled to stay warm as the pale gray darkness of day transformed into the cold blackness of night. Between the thunderclaps and the river's rampage the volume of noise was maddening. It was a long night and exhaustion succumbed to sleep.

The next morning the men struggled with the weight of wet tents and soggy bags of rice and beans and saddled up for the next push toward the river with the foreboding name. The Tigress waited for them just over the mountain. Bill shook the water from his brimmed hat and checked to see if his Marine Corps emblem was still intact. For some unexplained reason he felt good although he'd had a restless night and his body was aching from yesterday's ordeal. "Perhaps the sun will come out today," he thought out loud, as he stepped into the kneedeep mud to survey what lay ahead.

The tumultuous element of water had pulverized the mighty Andean Cordilleras into silt and mud. It was shocking. Everywhere he looked there was total destruction. Mighty cedro trees were angled precariously at their watersoaked root base. Across the playa cana brava stands along the Tutanangosa were flattened. Rainfall during the night had swollen the mighty river and she rolled in a deafening crescendo.

"At least we don't have the mosquitoes and those damn black bees to contend with," Bill said to his eager sidekick, Alfredo.

The words were barely spoken when the sun peaked out from behind gray clouds and once again they sucked it up and headed into the uncertainty of what lay before them.

They traveled for several hours in the warm sunlight before it disappeared behind the approaching black clouds that signaled more doom and gloom. Almost as if on cue the rains came as they reached a formidable swath of jungle roots and clumps of creeper vines that took the brunt of the razor sharp machetes. Jungle vines seemed to have group intelligence; they grabbed the men around their legs, neck and arms as they struggled through the trail less maze. It was the perfect setting for a snapped ankle, twisted knee or broken foot caused by the roots and vines that crawled unseen through the mud and foliage. Bill admitted openly that the thought of a broken leg was his greatest fear.

"How could you possibly carry a man out?" he mumbled.

The misery descending upon them appeared unnatural and bizarre. Their only consolation was in knowing there were no snakes to contend with because they only came out when there was warmth. Examining the turbulent skies, Bill knew that the reptile problem was only a remote possibility. The big cats and their constant search for food was another matter.

An even more deafening sound signaled their arrival at the Rio Tigress. They could hear the river before they actually saw it and when they did lay eyes on the Tigress their spirits plummeted. Danger lurked in the swirling brown current that was tumbling towards the Tutanangosa. Antonio quickly surveyed the turbulent river and ordered his machete wielders to begin searching for Cabuya, a tree fiber resistant to water and humidity that was used to make rope. The Shuara found the Cabuya and cut long sections for the men to hold on to as they crossed the river. Even with the lifeline it was impossible to stay upright in the swirling vortex of the strong current. They were turned back.

Antonio explored up stream and found an ideal crossing point. Rotten logs had washed downstream and were wedged together in a makeshift bridge. The logs offered footing and the Cabuya strung between the men offered them balance. One by one they began the slow and tedious crawl across the slick, moss-covered logs. Rushing water and jagged boulders were thirty feet below the fallen timbers and a slip would have resulted in broken bones or worse. Slowly and steadily the tired men crossed without mishap.

The Shuara established camp across the Tigress as an orange colored moon peeked through the canopied jungle above them. The men's spirits lifted when a rotten Acapu log, brittle and water-resistant, took to the flame of Bill's lighter. A lean-to constructed from Cana brava protected the fire from a light drizzle while the men enjoyed cooked beans and pork, their first hot meal in two days. They felt safe eating meat; there was no wind to carry the strong scent. The Indians were accustomed to alternate periods of food scarcity and bounty. When there was plenty, their instinct was to eat as much as possible to prepare for the days when food was meager. That night they gorged themselves.

Bill took a wardrobe inventory; three pair of socks and all were soaking wet. He meticulously laid out the socks near the fire then put on a dry sweatshirt that had been wrapped in plastic. He noticed that a young Shuara had no spare clothes and sat by the fire trying to get warm. Bill reached in his backpack, fished out a dry shirt and gave it to him. The boy, a deaf mute, smiled in appreciation. The gesture of goodwill didn't go unnoticed by the other Shuara. They smiled and nodded at the goodhearted Marine who had just crossed the line to become one of them. They knew that Alfredo despised them for what they were and that Oriolfo and Rodriquo were Canari and frightened to death of them. But this man who cleaned and oiled his weapons every night before he cleaned himself was no longer a colonial in there eyes, instead he was a warrior with a heart of gold.

One of the Shuara serenaded the crew with his reed flute while Antonio quietly wandered down to the river's edge, seemingly lost in thought; he sought his own private space. Unknown to the others, Antonio was tormented by his inability to understand what he was up against on the north bank of the Tutanangosa. In his previous expedition it was tigers and snakes and now unpredictable violent weather.

Later, he confessed to me that he had planned to take Achacho, his mentor, with him to the sacred waterfalls, take natema and seek answers but the hurry up expedition took precedence. Now he was stuck, wrestling in his mind over what to do. He couldn't turn the expedition back without Bill's approval and getting his approval required telling him about the dark magic interference. Bill had a sense of fairness about him, but he might think Antonio was not mentally fit to lead future expeditions. That would be a tragedy because, in Antonio's mind, he was the only person who might be able to work out a negotiable solution with his tormentor. The other problem bothering him was that Bill seemed to be under the influence of Alfredo's greed.

Antonio knew that on the next leg of the journey they would enter the corridor leading into the powerful but unseen world of discarnate energy that controlled the real world of fanged and long tooth predators that resided beyond the Tigress. He wondered if the wawek he feared would send in the night stalkers as it had done earlier or would it command the weather to continue its unmerciful attack against them? A lightening bolt followed by a deafening thunderclap brought Necta his answer.

The next morning they broke camp under crackling thunder and lightning then entered a cana brava thicket swaying to the pounding of the torrential rain. It was raining so hard they couldn't hear each other or see more than ten feet ahead. Bill said he was near the breaking point on several occasions because he couldn't see anything except sheets of water that stung his eyes. They were in tiger country and blind. In the rain and mist the jungle was colorless with only shades of gray in undetermined forms. The heavy carbine he was carrying was useless. If an attack came it would be quick and deadly. A bolt of lightning flashed and the shadows from beneath the trees suddenly appeared life-like and unnerving. He was on the edge.

On a break he told Alfredo, "Hell, we're totally at the mercy of Mother Nature and she's just toying with us." Alfredo didn't comment. He didn't have to because his face said it all. He was scared.

Everything looked alike as they slogged through mud that drained their stamina. Bill didn't know if the Jivaro were foolish or brave but they were the toughest S.O.B's he had ever met. He couldn't get over the fact that they never complained. They were so used to hardships that an expedition up the Tutanangosa for three dollars a day was nothing more than a strenuous walk for them. One thing Bill knew for certain was that these Jivaro were the cleanest bunch of guys he'd ever met. Each night they washed themselves, including their hair, and used green banana leaves to wrap their clothes in to keep them dry. But to try and keep up with them on the trail, forget about it! They had legs of steel and the lungs of a stallion.

That evening they reached the Santa Rosa River and through sheer determination managed to set up their permanent camp before nightfall arrived with a steady downpour that continued to sap their strength and confidence. After much difficulty a fire took hold and hot coffee was prepared with their usual meal of rice, beans, potatoes and soggy crackers. Those that were on the exploration detail preferred to stay away from pork; they did not want to tempt any jaguars along the trails the following day.

Under the dim light from a fog-shrouded lantern, Bill and Alfredo poured over Antonio's hand drawn map. Bill decided that the expeditionary force would split into three groups. One group, led by Antonio, consisted of Yaupi, Alfredo, and two of the carriers. The other group led by Bill would include Oriolfo and two Shuara. Rodriquo and the deaf mute would attend to camp. Antonio's group would explore the banks of the Santa Rosa and Wajai and Ikiam would take Bill to the 'Cave of the Tigre' as Wajai so aptly named it.

For five days the rain refused to let up as Antonio's group crisscrossed the banks of the Santa Rosa trying to locate the illusive rock, open in the middle and said to contain tres arrobas (75 pounds) of emerald and matrix. Constant rainfall made landslides a serious hazard as mountain slopes and the previously chopped trails vanished in swirling blankets of water and mist. The relentless rain was depressing; it soaked through ponchos and their wet clothes clung to their tired bodies.

The scary dark thickets of undergrowth were demoralizing even to the Shuara; long beards of moss that hung from the branches lent an air of edgy mystery to the savage land. Visibility was zero and the gnarled branches and limbs cast eerie images of strange forest creatures waiting to pounce on anyone that came near. To a man they regretted the day they had agreed to undertake such a miserable journey. The men had suffered greatly; all had sustained sprains, twisted ankles and ugly scrapes from the thorn bushes of the Amazon rain forest. Through all the mishaps and pain the quest continued. Antonio's group searched for what Richard Ireland referred to as the 'sultan's hat' while Bill and his group explored the damp dungeon that had been badly labeled as the cascabel.

When they reached the cave they unlocked their weapons and prepared for the worst, aiming the 44 carbine and the AR7s at the mouth of its opening. Oriolfo and Bill peered through their gun sights as Wajai and Ikiam screamed at the top of their lungs hoping to attract attention and to bring any tigre out in the open. After an insane few minutes it was obvious to them the tiger had taken up residence elsewhere. They would soon learn why.

With lanterns and flashlights blazing they entered the large opening that slowly narrowed to a pool of muddy, putrid water that was thigh deep. Hanging vines crawled with hairy brown spiders and slippery footing made exploration difficult in the cave that seemed to have no end. They sampled and explored what Bill called the "damp dungeon." It was a depressing place, a cavern of stale air, stagnant water, and something that caused Oriolfo to break out with dozens of blisters on his shoulders and neck.

The trips between the cave and camp were not uneventful. Each day was a challenge beginning with finding the previous day's trail. On the second day Bill's group got turned around in a cana dulce thicket and lost the trail when they stopped to investigate a large section of the sugar cane field that was swirled and broken by a tremendous power suggesting that two opposing forces had wrestled in mortal combat. The Shuara reasoned a tigre had attacked a large tapir, plentiful in the rain forests of South America and whose defense against the predators was only its keen sense of smell.

The frightening scene made Bill nervous; it confirmed that the rain did not stop the jaguar from hunting. On the last day in the cave his fears were magnified when Ikiam spotted fresh cat tracks on the trail. There were large, deep paw prints alongside the men's footprints from that morning. Bill judged, from the size of the prints, that the jaguar was large, maybe over two hundred pounds and had been stalking them, waiting for a straggler before it made its move. That evening Bill traded weapons with Antonio. Bill took the sawed off shotgun and gave the 44 magnum carbine to Necta.

Later that evening they mapped out their strategy for the coming days. The cave had been explored as far as possible and samples were taken from the dungeon's walls and floor. There were no green stones and no gold, only iron pirate, which was abundant. Finding the mythical cascabel was ruled out. The only lead to that puzzle was the cave they had just spent three days in had been eliminated for any future exploration. Antonio's group had explored the east bank of the Rio Santa Rosa and no large rock had been found. The men were in a quandary over what to do next.

Failure brought out hostilities; Bill and Alfredo got into a heated argument over where to search for the emeralds. Antonio remained silent, fulfilling his commitment to me, yet focused in his own world of mystical reality. Bill had been brainwashed by Adriano and me; our powerful argument that the emeralds had to be on the north bank remained his focus. He insisted that the two groups mesh into one expeditionary force and in the ensuing days they would explore the big river all the way up to the Ojal River. Alfredo refused to back down from his theory that the stones were on the south bank of the big river and that the Rio Cunquinza was the river that the Colombians named the Santa Rosa. Antonio realized that the arguments were going nowhere. He excused himself and called it a night so that he could go to his tent and think.

The minute the two men were alone Alfredo's voice dropped into his patented whispers that displayed paranoia and distrust of the Shuara. "Guillermo, esmeraldas aqui," Alfredo whispered as he pulled out a crumpled piece of paper and began to draw another map of the area. "We must explore the Cunquinza," he repeated over and over, locking eyes with Bill.

Alfredo explained his rationalization of why the stones were on the south bank and why the Colombians would have renamed the river. Antonio eavesdropped until the rain drowned out the muffled conversation.

Bill listened intently and saw merit in his passionate argument. He saw Alfredo as sly, tough and gutsy. "Through your veins runs the blood of a conquistador!" Bill said. Finally he told Alfredo to shut up and get some sleep; his argument was moot. It was impossible to cross the rampaging big river to explore the Cunquinza.

Between the Santa Rosa and the Ojal Rivers was an unexplored area of mountains and exhausting ridges that had to be chopped under, over and through. It was impossible to stay near the bank of the Tutanangosa because it meandered like a twisting snake. The only solution was to chop a trail about fifty-meters along the first ridgeline, which meant the men were constantly climbing or descending in gooey muck and rotten foliage. The roar of the river was their benchmark for distance. If its noise subsided then they knew they were too far inland and, when they could, cut the trail back toward the loud sound of the river. At times the jungle was so thick that if a man were to separate from the others and not be able to hear the river, he would be lost. That possibility weighed heavily on each of the men. There was no sun or night sky to get navigational bearings from, only the sound of the angry Tutanangosa.

Each day the men left base camp and crossed the Santa Rosa River on a makeshift bridge of timbers that the Shuara felled with their machetes. They crossed the whitewater twelve times and each crossing was another test of strength and courage on the rain soaked slippery logs. Only one mishap occurred: Bill slipped and was about to go over the edge head first into the jagged rocks below when Antonio grabbed him by the collar and hauled him to safety. The fall wrenched his knee and he was forced to return to camp for the day.

Antonio led the men on, hoping that when they returned to the previous trail chopped days earlier, it would still be intact. Some of the side trails had already vanished in the rapidly encroaching rain forest. As the days wore on lady luck stayed with them. They had not encountered any tigers, black bears or black panthers that, according to the Shuara, lived in this remote part of the forest. They saw large tracks in the soft mud and counted three sets of tracks on a previously cut trail that led towards the river. The predators had killed off most of the game in the region but this area was full of hundreds of beautiful multicolored birds.

When they finally made it up to the headwaters of the Ojal the air was much thinner and it was hard for the smokers to breath. Bill knew that any further exploration to look for a rock open in the middle was futile and called a halt to everything. Bill was rapidly reaching the end of the line. He scribbled the following in his notebook.

"I am physically beat with cramps in my legs, a twisted knee and cuts and bruises all over me. My body is constantly numb from the cold and endless rain that makes every joint ache to the point I think I am coming down with the flu. This scares the hell out of me because I can't afford to get sick in this place because every ounce of strength in needed to climb the next ridge."

Later he admitted, *"Mentally, I wasn't doing well because of the rain. It rained so hard and so long that you couldn't see anything and we were chopping, chopping all the time trying to make a trail. In the jungle everything looks the same and if you encounter a tiger you'd be lucky to get off one shot. And forget about the snakes! We didn't see any but we saw where they had killed birds. If you were bitten you wouldn't know it. Because you're in so much pain anyway the bite would just blend in and you'd think that you snagged another hanging vine or thorn. I know why they call the jungle the Green Hell. It wraps you up in a big, green, wet blanket.*

"If you wanted to torture a guy all you'd have to do is tie him to a tree and leave him alone for a few hours. When you returned he'd tell you anything you wanted to know beginning with the day he came out of his mother's womb. I never thought I would say this but this place is too rough for a white man. Damn, I'd given anything to have a radio with me and be able to listen to some music rather than the constant flow of Shuara and Spanish. I'm mentally drained from translating."

In the end it came down to one thing: futility. Striking camp they headed back down the north bank toward the settlement of Sucua. Between the Tigress and Victor Rivers they ran into a Shuara hunting party of men and women. They were a primitive looking group and it was hard to distinguish the men from the women because they all had powerful builds. Their arms and legs were solid muscle and every one had tribal war tattoos on their face, a feature that was permanent for life. Antonio talked to their leader but the rest of their group refused to have any eye contact.

Further down the trail the crew ran into soldier ants. Antonio took the team around them and said the ants were "muy bravo" (very brave) and that the rain brought them out of their hills. They were on the march and when the men stepped to close the ants attacked their boots in a driven frenzy.

Bill couldn't imagine falling off a ledge into an army of soldier ants, "Hell, they'd sting you to death before you could get away from them."

The tattered group reached Wakani, dried out near a roaring fire and feasted on baked corn and pork steaks under a star-studded universe. It was the first time they had been warm, dry and well fed in over a week. To a man they knew in their hearts that they hadn't failed in their mission. They had only been defeated. One battle does not determine the winner of a war and although this battle went to the Macas Forest there would be other battles and hopefully, in the next expedition, Mother Nature would stay neutral.

Somewhere up there in the womb of the green hell lay a fortune in emeralds that have been waiting over eighty years for someone with courage and daring to claim them and place them amongst his wealth. Bill and his band of explorers had given their all, and although they did not find the cascabel or the sultan's hat, they found within themselves an inner-strength to carry out their mission against unbelievable hardships. Most importantly they came back alive.

I was intently studying maps of the Oriente, wondering what had happened to Bill and his men when he walked into our house in Cuenca unannounced. My first reaction to his appearance was shock and then sorrow. He looked old and worn and had a dazed, hollow look in his eyes; like a chipmunk before a rattlesnake strikes. He appeared to be hypnotized but his eyes darted around the room searching for any sudden movement that might spell danger. His shirt was torn and his pants stained from mud and foliage. His speech was slurred and he had trouble staying focused. I suggested he get some sleep and that we could talk in the morning, but he refused and said he wanted to tell me what happened.

I handed him a cup of coffee and grabbed my tape recorder. My first question was what was the weather like? Bill answered in a slow southern drawl fighting to recall those memories he was trying to forget. For the next hour he rambled on with details of his frightening experience.

"Hoss, it rained everyday and night the minute we left Sucua. I think we saw the sun twice. Hoss, the nights were even worse. We didn't sleep. We were just waiting."

We continued to talk into the wee hours in between Bill's running to the bathroom with a bad case of diarrhea. His distant gaze and stunted speech was a sad ending to the aggressive beginning when we bid each other farewell that day another lifetime ago in Sucua.

The first night back Bill slept on the floor in a sleeping bag trying to get warm and fight off a high fever. After some tense moments when he hallucinated and spoke to some of his buddies in Korea, he finally settled down and entered a deep sleep. It was only then that I quietly entered his bedroom and removed the sawed off shotgun and the razor sharp machete poised for action next to his sleeping bag.

The next day Bill left Ecuador for the United States. The last thing he said to me as he boarded the flight was, "Hoss, if a man values his life he won't go there."

CHAPTER 12

RITE OF PASSAGE

Lee on trail to Wakani

After Bill said sayonara I retreated to bunker mentality and allowed my feelings to fuel the depression of failure. Had my luck gone south? Was there a sinister curse connected to Mejia's emeralds? It's strange what one remembers under circumstances such as these. Then, it dawned on me that every treasure had a price along with a different set of circumstances, and possible curse, attached to it.

Mejia couldn't return to his emeralds because he was an exiled invalid. According to the Indians my trips to Infiernilios ended in near disaster because the mountain spirits weren't pleased. I never returned to Infiernilios because of my curse: bitterness and anger. Now I felt as if I was being denied exploration of the north bank of the Tutanangosa because freakish weather prevented the steel cable from reaching the south bank. I questioned if all the treasures in this land had spirits or curses that dictated the fate of those who searched.

My mind raced back to the warning that Richard Ireland had given me in Phoenix. He had adamantly said, "I do feel that you need to be cautious, as there are other forces and powers against you. I'm speaking of physical powers and spirit forces against you. I want you to be careful! Move with wisdom and caution. Bodily danger involved, but you can circumvent that by using good wisdom."

But, Ireland also offered an uplifting footnote. "The native people themselves are going to be your greatest help. I see an old man...a very old man that could be of help to you." The native people Ireland referred to were probably Necta and his workers. But, I wondered, who was the old man that Ireland referenced?

Unknown to me, synchronicity had entered center stage. While I was encouraged by Ireland's words to continue the path of my search for Mejia's emeralds, Necta was also searching for answers. It was common knowledge by everyone around the settlement of Sucua that a dark force consumed some of the explorers that had gone up the river and never returned. People wondered what was responsible for the sun-bleached skeletons found by Jivaro hunting parties that had been picked clean by tigres and soldier ants? Was it the elements, the predators or a shaman?

Necta, who had questioned his uwisin power earlier, was on a vision quest to find answers through his mentor, Achacho. The old shaman was over one hundred years old and, in Jivaro tradition, was referred to as an unta, which meant "old" or "big" man. Revered for his longevity and respected as honest and magnanimous he was both an old and big man. His longevity was attributed to his supernatural powers; powers that some admired and others feared. Those that feared him believed he survived because he possessed the power to apply the death curse to anyone who opposed him or incurred his anger.

Achacho had been to war against their Jivaro archenemies, the Achuara, and taken his share of trophy heads. As he grew older he became an unta and advised other Kakaram (killers) prior to their attacks against the enemy. Upon their return the victorious Jivaro contacted him and asked that he be the wea, or master of ceremonies for their victory celebration. Early in his life he had known war, but age and wisdom had restored a gentle aspect to his soul. Mellowed by time he was now more interested in his garden than in war or curses. Necta found the old man in his hut near the Rio Upano and started a dialogue.

I had shaken my negative emotional state of mind and was slowly returning to my addiction to adventure. Perhaps I was influenced by Adriano's positive attitude and upbeat assessment that our bad luck was only temporary and Mejia's treasure was still within our grasp once the weather cleared. My change in attitude was fortified by a telegram from Necta requesting a meeting.

I added to the list of life's vexing problems another DC-3 flight through the corridor of uncertainty into the Jivaro settlement of Sucua. But this flight was different from all the others. It was gentle and drama free. Esus was waiting with two extra horses outside the dusty landing strip. Esus seemed genuinely disappointed when he learned that Adriano could not leave Cuenca. For Esus to show emotion was totally out of character and a reaction that surprised me. He asked if Doctor Vintimilla was ill and I informed him that Adriano was fine, but family business demanded he stay in Cuenca.

A few minutes passed and he once again asked if the doctor was ill. Once again I replied that Adriano was fine but could not make the journey because of family business. I got a little agitated when Esus again asked why my friend stayed in Cuenca. Then I got it. I realized that this was the way the Jivaro sought the truth. They ask the questions multiple times in various ways to see if the answer remained the same. He seemed satisfied that my third explanation matched the others.

Normally Esus exhibited one of two expressions, a smirk that resembled a smile or full out anger. There was a new expression; he was solemn and direct. "Antonio is waiting for you at Wakani, and we must leave quickly in order to be in the compound before nightfall."

Esus sternly commanded a cargo carrier to take the steel cable to Don Antonio Necta's house and then we headed toward the strong echo of that timeless river rampaging downstream.

Sometimes it's important to go with the flow, not knowing why or what the reason might be. I planned to deliver the steel cable and say hello to everyone, pass out a few gifts and be back in civilization by dinner time. Instead, I was sitting on a wooden saddle, astride a horse with a rough gate, with a backpack full of candy and cigarettes. I was also without a shotgun or food for a four-hour journey into tiger country. And to add more drama to my day was the underlying sense of urgency about this trip yet my guide, Esus, was distant and uncommunicative. I knew that there was something in the wind but passed it off. After all, everyone was entitled to a bad day.

For the first hour the trail was like a well-traveled dirt road that posed no serious problems, even with the small tributaries we crossed. It was a time to enjoy the sights, sounds, and smells of the abundant life force surrounding me. The new odors stimulated the sensory glands and about the time they became enjoyable they left, only to be replaced by a different smell that was even stronger than the previous. Occasionally pungent odors left my sense receptors reeling. The decay of rotting vegetation took center stage then quickly faded. Sweet odors filtered in and the cycle began all over again.

Startled by our incursion, a flock of blue parakeets sailed out of a tree covered with small flowers. Vivid colored Toucans held court in treetops, chattering and staring at the intruders below them. Always present was the powerful crashing sound from the Rio Tutanangosa.

Before long the smooth wide trail narrowed to a few feet across and then to inches when we entered a maze of hanging vines surrounded by grotesque trees with large gnarled feeder roots. It was as if nature, the artist, had run out of color and decided to paint in dark grays and ugly browns that assaulted my senses. How could this colorless world exist side by side with the exquisite canvas we had just crossed through?

Dismounting gave my backside a respite and my poor horse the opportunity to get his second wind. We led our horses ever so slowly, one foot at a time, making sure they did not hang a hoof in the creeper vines that Bill Wright had feared. The swamps were always the worst to travel through; I always felt hemmed in and captive to its ugliness and darkness. The Macas Forest seemed to know that and peppered our trail with these despondent passageways.

Navigating with caution through the swamp took some time but we encountered little resistance. Giant trees seemed relieved to have sunlight giving their roots a chance to retain strength and support. Our trail was still a river of mud with decaying vegetation and rotten tree trunks lying in shadowed areas. The drooping cana brava described by Bill was once again standing upright and swayed in the gentle breeze that cooled my sweaty body. It was a surreal experience to see the remnants of the storm's destruction blending in with nature's healing that was slowly regenerating the rainforest.

The swamp gave way to a steep incline and when we finally reached the top of the ridge we started a descent of several hundred feet in sloppy mud. We dismounted and dragged or pulled our reluctant horses until the cycle began all over again. My body wanted to scream from exhaustion. It was impossible for me to conceive how the previous expedition, with heavy loads of supplies and a monsoon battering their every movement, managed these ridges. I closed my eyes and allowed my imagination to come through; I could hear the men swearing and see Bill and his beleaguered crew struggling on these slippery and dangerous slopes. The ghosts of that expedition still haunted me.

Over the last ridge we entered a small valley, waded knee-deep in muck and slogged through swamp grass. I don't know how many times I stumbled. I was more concerned about my poor skinny horse that struggled with each step. We passed beautiful broadleaf plants, unusual trees and abundant cana brava. It seemed like the closer we got to Wakani the bigger the leaves and more dense the foliage. It was the perfect hunting ground for a predator and I couldn't imagine being on this trail after nightfall. Hell, I couldn't imagine being on this trail during daylight without any heavy firepower, but here I was, totally relying on Esus and his old twelve gauge shotgun that was undoubtedly loaded with birdshot.

I didn't know or understand why I wasn't frightened. I wasn't even all that jumpy, but I knew my guard was up and observation was survival. I was a little uncomfortable because I had forgone my number one rule for survival: better to be safe than sorry. Weapons and strategy were my safety nets and here I was taking a walk on the wild side without either. For some strange reason I felt protected by an unseen force as we pushed deeper into la selva. If Bill Wright had known of my predicament he would roll his eyes upward and say, "Hoss, it's about time you strapped on a shrink."

We crossed two more rivers and while our horses drank we paused for a smoke break. We were definitely in wild country. Thirty yards from where we stood near the riverbank a strange animal appeared and began to drink. The slight breeze was blowing away from the big brute and he had not smelled or seen us. It looked like a large pig with a long elephant-like snout.

"Grand Bestia," Esus whispered.

The giant Tapir was high on the predator's favorite food list and I hoped that he was not being stalked by his archenemy, the jaguar. When it finished drinking and looked our way it froze for a fleeting moment then bounded into the thick foliage. Esus smiled and told me that if it were not for our tight schedule we would have fresh meat for our evening meal.

I asked the small bow-legged man with the nasty limp why it was so important for me to meet Antonio at Wakani.

"It is what Don Antonio requested," he said nonchalantly. Without any additional input he said we must go, led his horse into the jungle and disappeared amid the thick greenery.

"So much for a real conversation and a valid reason for making this trip," I thought. We entered another clearing and spooked a large deer. It darted off into the brush, which startled a flock of birds nesting in treetops.

Around a bend in the trail, high above us in a large cedar tree sat two scarlet birds that the Shuara call Guayacans. They gazed down on us, cocking their heads from side to side in curiosity between cleaning each other with utmost precision and gentleness. Oh, to be a bird! That was the perfect solution to travel in this forest. No more ridges to climb or mud to slop through or rivers to ford. All a bird had to be cautious of were the tree vipers that lay in wait at the top of the branches on a hot summer day. I would trade the other dangers for that one I thought as I sucked in more air and began another grueling uphill climb.

We had been on the trail for hours when I saw my first reptile, a seven-foot brown snake that slithered through the decayed vegetation and disappeared toward the river. The snake was traveling on its journey, speeding along towards its unknown fate, much like me. I was amazed at the fluid motion and its quickness. He appeared and disappeared in a twinkling of an eye and had it not been for the movement on the jungle floor, I would have never seen it. I wondered what else might lay in wait, pacing us, watching us as we slowly navigated the rough trail to Wakani.

Half a kilometer down the trail my horse sensed danger, spooked and reared. At first he whinnied, then made a low guttural sound, and then his eyes widened with fear. He literally froze in place staring at some large rocks near the big river's bank. I followed his stare, which led me to a beautiful black and red reptile coiled on top of a boulder on the edge of a playa alongside the Tutanangosa. Ahead of me, Esus heard my horse's warning and turned back to check on us. He emerged from a thicket when his horse also detected the danger.

"Culebra es muy peligroso," Esus said unleashing his old shotgun.

Esus handed me the reins to his skittish animal and stealthily moved towards the ugly six-footer that was very much aware of his presence and moving into a position to defend or attack. Standing on the riverbank elevated approximately five feet above the snake he took aim at the angry menace that was now busy giving him a silent tongue-lashing. It was the snake's way of warning him not to come any closer while at the same time using its tongue to scent fear and purpose from an enemy. I held both horses' reins tightly knowing that the blast from the shotgun would cause the animals to react. The last thing I wanted was to spend the afternoon searching for Antonio's horses in a thick jungle or, even worse, having my arms yanked from their sockets by the animal's sudden recoil from the shotgun blast.

I braced myself, muscles taught. I waited...and waited... and waited. Esus was in no hurry but he was, in my perspective, getting dangerously close to the fanged killer. Then, as if he were guided by some strange tribal tradition, he lowered the weapon slightly and began speaking to the reptile in his native language. I had no idea what he was saying, but was fascinated by the exchange. Was he blessing the reptile before he killed it, like the traditional Apache did before they killed? Was he toying with the snake? I was mesmerized by the strange dance between man and serpent.

"Damn," I whispered out loud when Esus moved even closer, climbed down the embankment and was no more than ten feet away from the reptile. Is he crazy? Is he trying to prove his manhood? The Jivaro were without peers when it came to bravery but this was beyond courage and bordered on either insanity or stupidity. I wasn't sure which. For each step Esus took closer to the snake, it elevated its head and neck another six inches until they were face to face at eye level and only the length of the reptile separating them.

My nerves were shot, sweat trickled into my eyes, and I unconsciously held my breath. My mind raced back in time to that afternoon in Cuenca when Necta gave me a beginner's lesson on the culebras of the Oriente. What did he say about the snakes? Was it one of those that enjoyed persecuting a man? Was the venom hemotoxic or neurotoxic? Would it bury those long, ugly fangs into Esus and then coil nearby making sure he was dead before coming after the horses and me? I remembered the names, but none of the details about their killing pattern. Was it a two-minute or three-minute lethal snake?

I remained as calm as possible. Instinctively I searched out high ground as a defensive posture for the horses and me. I wanted to scream at Esus to shoot the son-of-a bitch, but I knew if I startled him or the snake with a loud outburst, he would lose his focus and that might be the trigger for the reptile to strike. I bit my tongue, gritted my teeth and squinted into the bright sunlight dancing off the river, waiting for the final outcome of the confrontation.

In a flash it was over. To this day, I cannot accurately describe in detail the following few seconds that left me confused, trying to sort out a foggy world of illusion versus reality. I remember that my horse flinched and I glanced at him for just a fleeting second and at that exact moment, a large, dark shadow appeared, momentarily defusing the sunlight. I heard an ear-piercing screeching sound that defied description. When I glanced back towards Esus, I thought I saw a large bird in flight, with the snake dangling from its beak. What kind of bird it was, I couldn't say, but it was large, so large that the wingspan momentarily blocked out the sun when it swooped down on its prey. It was as if the shadow of death prevailed on this strange mystical day.

Esus joined me; I was still in a semi-state of shock over his antics and the subsequent events. Had I just seen what I thought I had seen? My body was numb, my muscles tense, and the look frozen on my face caused Esus to burst into laughter.

"Senor Lee! Senor Lee!" He choked, wiping away tears of laughter. "You look like a man who saw a specter? Are you okay? Do you want some water? Do you want to rest before we go?"

Finding a few grams of remaining energy, I shook my head.

He replied, "Bien," and pried his horse's reins from my fingers.

They slipped silently into the darkened shadows on the trail. I followed along behind, baffled, replaying the events in my mind.

Down the trail we faced another river to ford and waded into the waist-deep, cold water, dragging the poor horses behind us. On the opposite bank we slumped against large rocks and massaged the numbness from our legs and feet. It was a good excuse for a much-needed break and it was the opportunity I had been awaiting for.

I handed Esus a coveted American cigarette and asked what kind of bird had flown off with the poisonous snake?

He lit up, took a deep drag, and responded, "Grande pajaro."

I knew it was large but, "What kind of bird?"

"The bird has no name," he said.

I tried again. "What is the species of the bird?"

"No se," he responded.

"Was it an eagle?" I asked.

"No se."

Fishing through my Spanish pocket dictionary, I could not find the translation for the word hawk and the only other large bird that came to mind was the condor and it carried the same meaning in both Spanish and English. It was also the name of a gold coin in Ecuador.

I pressed for an answer, "Was it a condor?"

"Es possible," he replied.

Trying a different approach, I told him that if it was a condor, then it was the first one I had seen in the jungle.

"No, Senor Lee, we see it all the time here," he said seriously.

I asked again, "Do you believe the bird was a condor?"

"Es possible," he replied angrily.

He was intentionally avoiding my questions about the mysterious bird. The Shuara tactics of asking the same question over and over to seek the truth was now turned against him and he was rebelling against the redundancy of my questions. But I felt there was more to it, something he was hiding. I decided to try a different approach with a different subject matter. I asked Esus why he did not shoot the reptile. Why did he move in so close to the deadly snake?

"It called my name when I was steadying the shotgun to shoot it," he said without hesitation. "It called my name and then said something I could not hear and when I asked what it said, it asked me to come closer so that I could hear the important information it wanted to tell me." Esus grinned and flashed that gold tooth smile of his, "Senor Lee, do you believe in spirits?"

His question caught me off guard. After a moment I told him I did believe in spirits and that he was the third person that had asked this question. The first was Dr. Vintimilla's mother, the second was Antonio, and he was the third.

"Good! You understand why I did not shoot. A forest spirit was using the culebra to convey an important message to me."

"What was the message?" I asked dumbfounded.

"Some of the words I did not hear and that is why I moved closer. Then I heard the spirit voice tell me that our journey to Wakani was without danger. All the predators had been warned not to create harm to us. We were being offered safe passage. The spirit voice said that the tapir we saw near the river was being stalked by a large tigre, but when we arrived it stopped the hunt and followed the instructions of the forest spirit. It went away. It was not permitted to disturb us. Another snake, the brown one, on the trail followed these instructions, too."

"What else did the snake...I mean, the forest spirit say?" I stammered in disbelief.

"Senor Lee, this is where the forest spirit becomes confusing. The reason I did not tell you these things back there on the trail is because I needed time to think out the true meaning."

He saw my confusion and knew he could not adequately describe his experience. He shrugged, "Colonials do not understand these things."

"Esus, I'm not a colonial. I grew up with Indio's. I respect their knowledge. I'm trying to understand, but I am confused. Why did the forest spirit speak through the serpent? Why did the bird capture the culebra? I don't understand. Are there many forest spirits? Did a bad forest spirit silence the good forest spirit? Why?"

"Senor Lee, you must speak with Antonio about these things. But, yes there are many spirits that live near the Rio Tungus."

"The Tungus?" I questioned.

"Si Senor Lee, the Tungus is the old Shuara name for the Tutanangosa. And, yes there is a constant battle of good and evil. It goes on every day, every minute with my people. This is the natural way of my people. You must speak to Antonio about these things. He will have the answers to your questions."

With that final statement our conversation concluded and we headed toward Wakani once more.

The sunlight filtered rays through the jungle canopy as we continued silently through the magnificent rain forest. Once again, I was experiencing the sights, sounds, tastes, and odors of a place that was ancient. It was strong and forceful, but somehow delicate and fragile at the same time. I realized that every one of my excursions into a jungle was with a focus, intent on reaching a destination. I had never taken the time to really absorb and be absorbed by its mysterious beauty.

I realized that this was what struck fear into the hearts of Bill, Alfredo and the others less than a month ago. I understood Bill's hesitance to return. They tried to penetrate the womb of the Macas Forest during a monsoon. This was a bright sunlit day and I was struggling to stay close to Esus through the debris and muddy remnants of that dangerous storm. I couldn't picture what it must have been like for them, but I was sure that Bill was nuts to try a treasure hunt in this place during that storm.

I caught myself in mid-thought and heaped a dose of reality on myself. Who's the nuttiest? Hell, I would win the crazy contest hands down. Bill had lethal weapons, he was prepared, he had safety in numbers, and Antonio, a man that I respected and trusted, was leading him. Here I was a month later, lacking all the safeguards, being led deep into the jungle by a man taking instructions from a deadly serpent!

Dry season, wet season, it was practically all the same up here on the north bank as we edged closer to the mystical Wakani. Even on a bright sunny day everything in the darkened shadows of the trail was wet and clammy. Did it ever dry out? I continued to search for any sudden movement along the edge of the trail, despite our supposed guarantee of safe passage. There was no question in my mind that Esus believed he was in touch with a forest guardian that was communicating through a venomous reptile. It was his truth, his reality, but was it mine? I thought about the old prospector in the Hotel Humboldt that cautioned me about going into the Oriente. "Never believe anything you hear and only half of what you see. The Oriente has a way of twisting your reality and beliefs."

According to Jivaro traditions the spirits ruled the forest and were seen on a regular basis when under the influence of natema. This was a normal occurrence for those that believed in the power of the unknown while in a trance state of mind. But, neither Esus nor I had taken drugs. We were operating with full faculties when my horse sensed the snake. It was comforting to know that my horse sensed the serpent before either of us saw it. I saw Esus move close to the snake. I saw the bird swoop in and take the snake away. At least I thought I saw it. I was comfortable with what I had seen but had reservations, questions that revolved around the communication between the forest spirit represented as a snake and Esus. I saw Esus' lips move after he lowered the shotgun; the reptile was probing with its tongue, but to say they were having a conversation was a bit of a stretch for me.

I wrestled with a creeping paranoia that nudged into my thoughts. Did Esus really have telepathic communication with an entity or—this was really bothering me—was he under the influence of insanity brought on perhaps by taking natema over the years? Was Antonio waiting for me at Wakani, or was this a ploy to frighten me, or worse yet, remove me? Would my head become another trophy for the headhunter? His statement about his people wrestling daily with good and evil was not reassuring. The Jivaro culture was based more on war, revenge, death, and destruction than anything resembling what an Anglo would consider normal and this had been their tradition since time began. Maybe I was a colonial? I was wrestling with dual realities.

Esus' truth was steeped in tribal tradition and he believed that the supernatural was a force to be reckoned with, exactly what he had been taught from the time he first received his arutam soul. My resistance was based on my refusal to believe that he was conversing with a serpent. Then I realized this concept was not really that foreign to me. As a youngster in bible schools the first thing I was taught to believe was another story about a serpent that spoke to Eve in the Garden of Eden. If I believed that snake story because I was told it was true, why was I questioning Esus' truth and something I witnessed?

I continued to struggle with the Amazon trail and my own jangled thoughts when it suddenly hit me that truth is simply one's own perception of something. As a youngster, my grandparents were immersed in the dogma of a stringent fundamental religion that influenced me. My grandmother used to say we were God's chosen ones and she and my grandfather believed this to be the truth. My young and malleable mind was steered into their religious beliefs and then brainwashed into submission where it remained for many years in obedience to their belief. So many people were trainable, I thought, doomed to follow a path chosen for them rather than their own. I was lucky, curious, and strong willed, determined to find a life suited to my desire for adventure. I couldn't decide whether my friend Esus was fortunate or unfortunate because he never had the opportunity to leave that group controlled culture that made him believe in the supernatural world of the denizens of the forest.

Was I judging him over something I didn't understand and had never been educated in? Had he been as confused as me over the day's strange events? I remembered that the God concept for the Jivaro was encompassed in the supernatural and the spirit world. Their idea of an omnipresent being was identified as the "earth mother", Nunui, who was responsible for crop production, protection, and seen in dreams and hallucinogen-induced visions dancing in the gardens after dark.

The only other omnipotent personage that the Jivaro revered was the legendary Tsuni, the mythological first shaman. Tsuni was not so much a deity as he was one of their creations. They believed that the energy of Tsuni resided within the shaman with the most power, which made it imperative for each uwisin to accumulate power. No wonder the men in black, so named by the Indians, who manned the many Catholic missions in Jivaro territory had strived to save so many souls. To them the Jivaro must have been desperate people in need of a Higher Being that could grant them eternal life because the Jivaro beliefs didn't recognize forgiveness and salvation.

Instead of a Divine Being their existence was orchestrated through their constant usage of N, N-dimethyltryptamine, or DMT, a powerful psychedelic found in ayahuasca, the sacred vine the Jivaro called natema. Young children were given the powerful drugs to help them find the true way and from the time of puberty they experienced the supernatural realm that was an amalgamation of myth, mystery, and reality. This truth they readily accepted as the only way of life. Even their dogs were given the drug in order to make them more proficient hunters and guardians.

Then I thought of friends back in the States who were experimenting with LSD and peyote as a means of better understanding themselves and the world around them. Were they much different than the Jivaro? Probably not, but the Jivaro had mastered the art of communing with the unseen, when most from the civilized world were taking baby steps.

I realized that every man is entitled to his own beliefs and my judgment was replaced with a knowing sensation that caused me to laugh outwardly at my uneducated assumptions. My revelation seemed to stir up a wispy breeze that rustled the leaves and cooled my forehead. I took it as a good omen and smiled up to the heavens and spirits of the Oriente. At that moment we broke into a wide clearing. We had arrived at the place that meant "soul" or "spirit" or "spirit bird" Necta's home, Wakani.

Nestled at the junction where the Rio Tunanza joined the Tutanangosa, the finca (ranch) that was named after a spirit helper bird sat on a bluff overlooking the Tungus. The land was cleared from the ever-encroaching rainforest and I knew it was a labor of necessity and love that entailed countless man-hours of backbreaking work. There were two structures side by side that seemed the complete opposite of each other. The large rectangular building was purely colonial in construction with clapboard siding and a wooden roof. Next to this building was a wood and bamboo lashed sleeping area that was elevated a good ten feet above the ground for protection against predators.

Wakani was a safe-haven amid a sea of green that stretched as far as the eye could see. It reminded me of Necta, a living dichotomy captured in two worlds, the colonial and the Shuara, and longing for the third world of the unseen. Across the Tungus, the old name for the Rio Tutanangosa was a primitive world known only as the south bank, a far more formidable opponent than what we had just come through. I wondered if its reality was real or the unseen.

It was late afternoon and the spirit residence was drowned in sound from the roar of the legendary river, which prevented Esus from calling out to let his father-in-law know we had arrived. My attention was drawn to the garden where Antonio Necta was stooped over with his back to us. Sensing our presence he stood and turned in our direction, shaded his eyes, wiped the sweat from his forehead, and then came running to greet us.

He was elated to see me although disappointed to learn that Adriano had stayed in Cuenca. He led me to the rectangle building while Esus filled him in on our journey. I couldn't wait for the opportunity to discuss the snake events with Necta and find out from him what he thought the strange bird was that silenced the forest guardian.

I was ushered into the non-elevated structure and to my surprise the tents and food supplies from the previous expedition were neatly packed in a corner of the room. They were perfectly arranged and cleaned of all mud; even the cans of food remained intact with no rust along the edges. It was as if they were waiting for me to come and claim them. I thanked my friend for taking good care of the equipment and supplies but also scolded him in a gentle way for not eating the food. He responded in a soft tone and a humble smile and said the food was not his to eat.

I scanned the room and noticed that it only contained a cooking hearth, fireplace, and elevated hardwood bench that ran the length of two sides of the room and I was sure it would feel like velvet compared to the saddle I had just left. Removing my mud encrusted boots I noticed that beneath the seating area was open space that left me feeling vulnerable to any creature that might slither through at any given moment. On the large open hearth sat two pots simmering on the low burning fire. The room was filled with the soothing aroma of chicken and yucca that accelerated my hunger pains. It had been a long day and I had only eaten hard candy and a few strands of jerky.

Necta saw my interest in the food and quickly pointed out that we were having a special meal this evening. Since this was my first visit to Wakani he had selected one of his finest chickens for dinner. He patted my shoulder and told me that after dinner we would talk, but first I had to wash. He handed me two small pieces of cloth and pointed in the direction of the river below. The setting sun appeared blood red as I made the long walk down the bluff through a field of short green grass that served as a pasture for grazing animals, although none were present. Antonio Necta and his vaqueros had cut back the jungle for over a hundred yards to where the river disappeared in a bend behind heavy vegetation.

It felt safe here at his ranch. But, a complete opposite lay less than thirty yards across the Tungus, on the south bank. It was an ominous looking place, a primitive and pristine tangle of rainforest. There were large fluted columns of giant trees and under these sentinels of the forest grew smaller trees and under them the broad leaf plants and ferns that began at the water's edge and ran into infinity. Less than a mile from where I stood it would be impossible to distinguish the difference between the south and north banks as they merged into one massive trail-less rainforest separated only by a meandering river named the Tungus by the Jivaro and the Tutanangosa by the colonials.

The Tungus was rolling, showing off its white rapids, but I found an inlet pool and large rock that offered steady footing. The water was cold but invigorating and the continuous crashing sound of whitewater careening into large boulders I likened to a form of peaceful anger, if there was such a thing. It was mesmerizing to realize that after all those months of research into Mejia's lost emeralds and those countless hours of tracing this river on old maps, I was actually standing on its legendary bank.

I knew that the cascarillas went to sleep every night being serenaded by the same sound that was now engulfing me. The thought brought chills down my spine. In the middle of my mental imaging of a cascarilla's life, the words of Crespi slammed my senses. "Beware of the boas in the Tutanangosa. When the sun sets they come out to wait for their prey to come to drink." I climbed the bank and headed for my dinner before I became someone else's.

I was surprised to see a very old Jivaro with Necta. He was stooped in a gentle way with black piercing eyes and matted white hair that was shoulder length. He was dressed in the typical Jivaro attire and wore sandals made from an animal hide. I was quickly introduced to Achacho and was told that, as a young boy he was present when the Colombians arrived in this area. Necta explained that Achacho did not speak Spanish and asked if it would be all right if he translated our conversations to him in his native tongue. Then, with a sly gleam in his eye, he added that Achacho might be willing to share some of his knowledge with us concerning the clues in the last will and testament. But first we would eat and afterwards we would talk.

My mind was racing. I was sitting with someone who was present when Mejia was alive and probably met him or at least had seen him. Then I remembered what Richard Ireland, the psychic in Phoenix, had told me at the end of the private session. "I see an old man…a very old man that could be of help to you. Go with God!"

Was this the man Ireland envisioned? But, Ireland had used the phrase; "could be of help" rather than, "would be of help." I wondered what would make the difference?

The old man was quietly studying me in a very non-intrusive, non-confrontational way. I felt like he was seeing me from the inside out, understanding the total man rather than the exterior facade. There was a subtle reverence and respect that emanated from the two men as they spoke in their native dialect. This had to be Necta's mentor, the wise old uwisin referred to when he explained to me that he was a pener uwisin and still being guided by an old shaman.

The brief information Necta shared in Cuenca had spurred me to do some research on the Jivaro at the University. I found the information extremely fascinating, but would never have believed that shamanism would apply to my expedition to the Sucua area. I assumed that what I read was about a tribe of Jivaro deep in the Amazon. I guess I had rationalized that Sucua was more colonial as the Church had missions all over the place.

The Ecuadorian colonists were bringing about changes and civilization was coming into the Upano Valley. Air transport made it possible and profitable to ship beef into the cities. Sucua was slowly becoming a commerce center. A bridge over the Rio Negro near Mendez was planned and the new road from Cuenca to the Oriente would be completed soon. Everything on the surface seemed normal to any outsider coming into the area.

I drank the cup of chicken broth knowing that I was a stranger invited into the private world of two respected and sought after shaman. It reminded me of the time as a youngster I had accidentally stumbled on the Apache Kid's teepee on the San Carlos Reservation and made friends with the old man. He had taught me many things about the desert and respecting nature and finding peace within. I felt like this was another round of those lessons, but I wasn't sure. I only knew I was comfortable here. Manioc beer was passed around. That was the beacon that heralded the beginning of a conversation that I was sure would last long into the night.

CHAPTER 13

HEARTFIRE

The clearing where Singusha's house once stood

I expected an exchange of information but after the first few questions it felt more like an inquisition was underway. At first the questions were centered on my feelings of working with the Shuara. I answered honestly; I had no problems working with the Jivaro. Necta visibly cringed while Achacho remained quietly observant. Puzzled, I asked what I had said wrong.

"We do not care for the name Jivaro. The colonials gave us that name. It means savage. We prefer our given name, Shuara. Shuara means man, men or people."

I felt like I was definitely off to a bad start. My elation of sitting with two great uwisins slowly dissipated with each direct question about my desire to find the green stones of the Colombian. I sat on the floor. There was no breeze, no circulation and the scent of the evening jungle was heavy and acrid. I tried to understand the mixture of Spanish and Shuara and maintain eye contact. I explained my intent and they listened patiently before asking the same question in a slightly different way.

In my previous meetings with Necta he had been more animated, filled with the wonder and joy of life. But, this evening he was a different person and seemed preoccupied with learning my truth. His presence conveyed strength and purpose and his speech was slow and meticulous as not to allow any room for error. Many of the questions came from Achacho and Necta had to translate to Spanish. With my Spanish dictionary in hand I was given the necessary time to seek out the correct meanings of any words that I failed to understand. I had the dictionary open often and knew they were carefully monitoring my actions and expressions.

Their questions were direct, which I was comfortable with and I kept my answers as brief and succinct as possible. When Necta asked if I trusted Alfredo, I got an empty feeling in my stomach; I realized that he didn't trust him, but I wasn't certain why. I looked him in the eye, took a deep breath, and told him honestly that I wasn't sure if I did trust him. I knew Alfredo was an opportunist and I knew he had inspired Bill to go up river in weather that I refused to attempt.

Necta translated to Achacho, then turned to me and asked why the Spaniard did not respect the Shuara. That left me in a quandary because I didn't know. I didn't want to seem shallow and I knew I couldn't defend a man I had little respect for. I shook my head and told them that I wished I could answer, but I could not. I did not have an answer. I was irritated, but not with the Shuara; I was irritated by Alfredo's actions. What had Alfredo done to compel their lack of trust? It was quite obvious that one or both of the uwisin had detected deception in him.

Then he asked about Adriano. Were his intentions pure? Without the aid of a dictionary my answer was direct and to the point. I wasn't angry, but I was firm. I punctuated my answer with a question directed to Esus and Necta. "How many homes have you been invited into in Cuenca? How many necklaces has your daughter, Maria, received from Cuencanos?"

I was being defiant and knew that I had violated Shuara protocol by asking a question off a question but I didn't give a damn. I felt fairness was two-way communication not one-way interrogations. Esus and Necta got the point as both men nodded, somewhat embarrassed while the old uwisin continued to study everyone's actions and reactions.

Something was wrong, but I couldn't put my finger on it. What happened on Bill's expedition that would lead to this? My mind was filled with questions but I wasn't certain how to broach them without offending the Shuara. Necta changed the subject, got up from the bench and went to the other house.

While he was gone Achacho and I locked eyes; neither of us faltered. It reminded me of my youthful days growing up on the reservation and the times the Apache kids and I would have stare down contests. The first one who blinked lost. My intensity was countered by the old uwisins' peacefulness. Just about the time I was ready to blink, Necta returned and threw another log on the fire. As the flame grew brighter, Achacho grunted some words in his direction then Antonio turned to me and in halting Spanish and asked why I wanted the green stones.

I sipped the warm beer and tried to gather my thoughts so that I could answer honestly, even though I wasn't completely certain what the answer was. I explained that the stones had great value in my world. If Adriano or I found the stones, we would not keep all of the wealth for ourselves. The Shuara would also benefit from it. There would be enough wealth that it could be shared. They could build schools for education, plant bigger and better crops, and live easier lives.

Both of the men were boring into my soul with their intense stares. Necta started to translate but I interrupted him and continued to explain my desire to find the green stones. I didn't know if they would understand but I told them that I was fascinated by the mystery of Mejia and the riddle he left. I wanted to understand the Colombian that left such a tempting trail to explore. I didn't want to be like the other colonials who had come to Shuara land, exploited it for personal gain without consideration for the Shuara, and vanished once they had taken all they could.

The uwisin studied my movement, my eyes, my entire being. I felt as if I were speaking words they did not hear because they were so involved with my exterior reactions and my truth that came from some place deep in my soul. Again Achacho grunted something to Necta. Antonio reached in his chest pocket and removed an object, which I could not see. He silently studied it, returned it to its hiding place, and then asked what I would do if I found the green stones. The Shuara were demonstrating their famous 'ask the question again in a different way and see if you get the same answer' techniques to find the truth. I sat straighter and looked Achacho straight in the eye. I replied that once the stones were found, Adriano and I would take them to my country and sell them. Then I would return with money and give the Shuara their share in the profits. I explained how this process could be repeated until other arrangements could be made.

I knew that the subject of greed was in the back of their minds after the Alfredo incident. I had asked myself before if I was greedy. I really didn't think I was, but what happens to an old prospector who hits pay dirt? A little won't do. Once he'd had a taste, like a drug, he wanted more and more. I decided to diffuse the issue early on by placing Necta in charge of my greed watch.

Necta translated to the unta. No one said a word and I began to wonder if my answers had offended the old man. I wondered if I had said too much or if I had not said enough. Did they want more of an explanation? Did I confuse them when I said I was intrigued by the mystery? Necta interrupted my thoughts, offered me a fresh cup of manioc beer, and asked about my early life, my father and what kind of man he was? The question threw me; I expected more dialogue on the emeralds. I explained that I never knew my father, that he abandoned my mother and me when I was very young. Necta seemed sad. I told him that my grandparents raised me and he seemed relieved when I explained that my grandfather; a hard working man, replaced my absent father and became my mentor for strength and guidance.

I shared that I had grown up on the Apache nation and gave them a brief history of the Apaches and their fight against the colonials. The Shuara identified with my story. I related that my early friends were young Apaches and that I had a special place in my heart for those people. Necta became quite animated when he translated the details to Achacho. The old man listened carefully and searched my soul with his deep black eyes then nodded his approval. It was odd, but everything, even the heavy, moist air of the jungle, seemed relieved and more relaxed. Just as quickly as it had begun, the interrogation ended.

Necta and Achacho entered into a lengthy conversation that only a Shuara would understand. I took a sip of beer as the lengthy exchange continued. It went on for quite sometime; Necta did most of the talking and Achacho listened. Occasionally Achacho would offer a word or two, a grunt or several nods but it wasn't the nods and grunts that had my attention, it was the language of the Shuara. It had a hauntingly familiar and comforting sound. What made it even more special was to know that a mystery surrounded the language. Anthropologists didn't know the origin of these proud people or how long they had lived in the Amazon rain forest or from where their language was derived. The only thing that was known was that the Shuara had kicked ass with the Incas and Spaniards and still remained unconquered in spirit. Their past was as mysterious as their present.

The gathering at the spirit house was an amalgamation of mystery, enlightenment, and strangeness. I was weary from the long day and the manioc beer and Antonio must have seen it. He paused in his discussion with Achacho and went to the corner of room and fished out two sleeping bags and air mattresses. He made our beds for the evening on the floor next to the hearth and then Necta said Achacho was tired and was leaving to sleep in the other house. I assumed that would give him privacy and time to explore my truth.

After the old unta left, Necta told me the story of when Achacho was a young boy and first saw the Colombians when they came to harvest the quinine. Achacho had known Singusha and told me how the uwisin befriended the man called El Sapo. Mejia was referred to as El Sapo (the frog) by his co-workers because he was small, had droopy shoulders, heavy jowls, and resembled a frog. Regardless of his nickname, Singusha saw beyond what the others made fun of; a good man that respected the Shuara and took time and interest in their traditions. Antonio smiled when he talked about the time the young Achacho snuck down to the water's edge and hid in the jungle to spy on the Colombian camp across the Tungus.

"Tomorrow if Achacho has strength he said he would take us to the loma (hill) where Singusha lived. From there we will be able to see the ruins of the Tambo Santa Rosa. But, now you must sleep and rest."

He left and went to the house on stilts. I sat staring into the flames from the fire.

Before long Esus bedded down a few feet away, went into a sound asleep and began to snore. I adjusted my sleeping bag, rolled over on one elbow and with my hand cushioning the side of my head and stared at the red-hot embers glowing in the hearth. I took a deep drag on a cigarette and heard Necta's words replay in my mind, "Achacho would sneak down to the water's edge and spy on the Colombian's camp across the Tungus."

My nemesis, Alfredo, the man the Shuara did not trust, had been right all along. Now I understood why Necta wanted steel cable to cross the Tungus. I felt comfortable in these humble surroundings that were so rich in other ways. I rolled up a wool blanket and laid back on it watching shadows from the fireplace dance around the room. Were these spirits, too, I wondered before I succumbed to sleep.

In the morning we gathered again for a meager breakfast of fruit, packed some food and water in the backpacks Bill had left in Necta's care, and departed Wakani up river. The sun was slowly burning off the spooky mist that surrounded Necta's compound but still hung thickly in the dense jungle. Achacho may have been 100 plus in years but he moved like a spry sixty-year-old as we made our way through heavy growth before we passed a clearing on a bluff over looking the Rio Tungus that had been renamed the Tutanangosa

Something about that clearing nudged my senses. It was magnetic. My skin tingled; my eyes darted about searching for something. I felt as if a powerful and unseen force unknown to my understanding was studying me. Was this crossover of energy attempting to get my attention? If so, it had been effective. The energy paradigm had made a strange connection into my consciousness. I stood dumfounded staring into an empty grassy area until Achacho stopped and with one hand on Necta's shoulder to steady himself, he pointed across the river. That was the location of the Colombian encampment named Tambo Santa Rosa. Then Achacho stepped back and with both hands stretched outward said that we were standing on the very spot where Singusha had lived.

"The Colombian, El Sapo, came across the river to eat and drink with Singusha until he left to fight in a battle between warring colonials," Achacho said.

With the return of my senses my imagination spun out of control as I stood on the wind swept knoll that was the foundation of the old dwelling that had been reclaimed by the rain forest decades ago. What had the old shaman, Singusha, been like? Was he really vengeful or just protective? Was he afraid of loosing this raw, powerful land and the spirits that inhabited it to men who did not understand or respect it? What did he think of the savage ways of the Colombians, their unusual rituals and forms of worship? Had Singusha's energy tapped into my consciousness?

I looked across the river to the grassy clearing. This was not a gentle land; I could only guess the hardships and loneliness of the Colombian quinine collectors as they carved out a living in this fierce terrain so far from their loved ones and homes. For a dollar a week, or less, they risked their lives on a daily basis, tempting the elements of nature, avoiding warring tribes and sickness to save the lives of others with the quinine they harvested for shipment to Europe. And only one man, Mejia, out of the entire crew had struck it rich, only to see his wealth slip away in an ironic twist when he got caught in the battle of Guayaquil, fighting Adriano's relative, General Vintimilla, where he lost his leg and eventually his Ecuadorian visa. It was an exciting, but also a sad moment for me. I gazed into the green, wet, blanket across the Tungus and wondered if it was hiding the ghost of Mejia.

Achacho studied me carefully. He was everything that Ireland had predicted. The old man was helping me beyond my wildest dreams; even to the point of saying that there was not one Santa Rosa River but three. There was the Santa Rosa River below the Ojal and off of it was a small stream named the Little Santa Rosa.

"Arantchiiti," (near there) he said with Antonio translating. Then he shaded his eyes with one hand and pointed across the river with the other. "Arantchiiti," he said again.

We all squinted and looked in the direction of a small river he said was the Cunquinza that flowed next to what used to be the Tambo Santa Rosa. Necta translated, "The Colombians came and named it the Santa Rosa, after one of their spirits. This was where we must search, but first, we must get permission for you to go there."

I readily agreed and asked where we had to go and get permission? Achacho said he was tiring and we must return to Wakani.

I was careful to note the different landmarks with particular attention given to where we were in relationship to the rivers that fed into the Tungus. The Tunanza was the river gateway into Wakani but the next major river had me confused. I wished I had my maps and was more prepared. I asked Antonio the name of the next major river.

He pointed upstream and said, "The Victor".

Damn! Ireland was right again! He had told me "...It has to do with Victor, all the way down to like...Teranzo or something like that..."

The Tambo Santa Rosa and the Rio Cunguinza were located between these two rivers but on the opposite side of the Tungus. How did Richard Ireland know that an old man would help me? How did he know that the Cunquinza River lay between the Tunanza and the Victor? He was so accurate that he even knew the Victor was the upper river when he said, "all the way down to the Tunanza".

Dr. Richard Ireland was my first experience with someone professing to have psychic powers and I was somewhat dubious of his nightclub act in the beginning but as the evening wore on his methods and information had caught my interest. By the time he came to me I was nibbling on his hook and then he reeled me in by naming the two rivers that only a handful of people knew. Now the information he had given me about the future and the past had come true. I had been accepted and helped by an old man, Achacho, who had given me the location of Tambo Santa Rosa and two major rivers bearing the legendary name Santa Rosa.

Ireland's accurate information left me with more than a favorable impression for those that claim these kinds of unique gifts. My thoughts went back to that evening in Phoenix during my private reading when he said, "I do feel that you need to be cautious as there are other forces and powers against you." If I only knew what those opposing forces and powers were I could prepare for them. Had one just tapped on my conscious mind while I stood on the grassy knoll that once belonged to Singusha?

Achacho and Necta sat in the grass quietly talking. Soon there was only the sound of the wind gently rustling the leaves and whipping up an occasional white water spout from the Tungus below. It was difficult to control my excitement after three years of speculation about the location Tambo Santa Rosa. The first piece in Mejia's puzzle had been found.

I took a long deep breath and absorbed the details from this experience. I had sought the help of a psychic in Phoenix, but

Necta was his own psychic. We each had our proof that an unseen world existed, a world that had information that we couldn't read from a book or hear in a conversation. There was so much I had to learn about my world and about his and, at this moment I was torn between digging deeper into supernatural happenings or remaining engrossed in the presence of a world that existed long ago. Achacho stood up, arched his back and strode off toward Wakani. That was our cue to leave. We walked back single file, in silence, each lost in our own thoughts.

Esus was waiting with the horses when we arrived back at Wakani. It was a tiresome haul and he was eager to return to Sucua. I took Achacho's hand and thanked him for his assistance; I thought I detected a faint smile cross his lips. Saying goodbye to Necta was difficult; he had shared a part of his life that was private and spiritual and I had shared part of myself with him. He knew our paths would cross again. I gripped his hand and arm tightly; my trip to Wakani had been successful beyond my wildest dreams with the exception of the mystery surrounding the big bird and the venomous snake, which, in parting, I asked about again.

At first he didn't say anything. Then he answered that he must get permission to build the cable bridge across the Tungus, which was an obvious avoidance of the subject. I smiled. I realized that each time I asked him to explain the forest spirits he changed the subject matter. My curiosity was a question whose answer lay in the future. I would just have to be content that he would receive approval to build a bridge across the Tungus.

I felt reborn. I had touched something mystical and magical within my being while I was at Wakani. Necta spoke of my "heartfire" and said that it had been awakened. He said a person's heartfire is the window into their soul and that a pener uwisin could "see" the heartfire that determined one's purity of thought and deed. Was he talking about my body's aura? Perhaps in time I would understand but now I was anxious to see what awaited me in civilization. What I found was the flip side of the Oriente's magic.

Dark forces marshaled under the guise of doing business in a country with antiquated laws involving foreign capital. Adriano informed me that we must abandon the Ecuadorian Corporation. He was exasperated by the ongoing comedy of errors that masqueraded as a legitimate international banking system and admitted if this was allowed to continue his country would never leave stone age economics.

I was saddened but also relieved. My struggle with business protocol had cost me valuable time: thirty-six days awaiting a cashier's check and most importantly a weather opportunity for an expedition. It was back to the balls to the wall approach that was more suited with my rebellious nature. Bob Olson would cringe if he knew, but I was headstrong and made myself a promise that my future expeditions would be self financed and void of any stringent rules or strings attached.

With the bad came the good and now it was time to bring Adriano up to date. "Dr. Richard Ireland may have solved one of Mejia's riddles, *like making a relation*, but now we had a more perplexing riddle." I paused to gain a degree of clarity and sort out my thoughts.

"Adriano we have a set of twins connected to just about everything we've been searching for. Case in point. Achacho told Antonio the original Tambo Santa Rosa burned down and a new camp was established further west. Now we have Tambo Santa Rosas on each bank of the Tutanangosa. We have the original Tambo Santa Rosa on the south bank and the second on the north bank of the Tutanangosa. Achacho said that when he was a young boy he spied on the Colombians across the Tutanangosa near Singusha's house that was located on the south bank of the Tutanangosa. In my opinion this is the Tambo Santa Rosa that Mejia spoke of.

"Achacho said the Colombians named the Cunquinza, the Santa Rosa. Later, the Colombians established a second camp further up the Tutanangosa and named a tributary flowing into it the Santa Rosa. Now we have our second set of twins. This is the one indicated on the army map that we have explored in the past.

If we are to believe that Mejia was telling the truth and we are to believe Achacho, then the only Santa Rosa River left to explore is its twin the Rio Cunquinza on the south bank of the Tutanangosa near the original Tambo."

Adriano chimed in. "Lee, amazingly, there is a third set of twins. They are the cascabels. Antonio recently told me he found out about this through an old Jivaro named Yangura, the son-in-law of Achacho who knows of two rivers named cascabel by the Shuara, because of the many small waterfalls in the tributaries. They produce the ca-ca-ca-ca sound of the snake and the tree pod that carried the name cascabel. The first one was renamed the Santa Rosa River by the Cascarillas and it's beneath the Ojal River. This is the one Bill explored unsuccessfully in '69'. It lies on the north side of the Tutanangosa River. "

Stunned by the information it became necessary for me to call a time out and think about the riddle that was now morphing into multiple rivers, camps and cascabels. Notwithstanding, was the major river that shared two names. The old Shuara knew the Rio Tutanangosa as the Tungus. My God, if all of this was true then it was no wonder the Mejia family and all the outsiders, including us, were batting zero in the attempt to locate the emeralds. Even the Shuara living in the area must be confused as to what Tambo, Santa Rosa River and cascabel they should explore? The Colombians definitely had a fixation with the name Santa Rosa and the Shuara weren't far behind them with their usage of cascabel.

How clever of Mejia to use symbols as his clues. In the beginning we were convinced that a waterfall might be the cascabel. When Mejia wrote, *"the entrance is by the cascabel,"* we immediately assumed a waterfall was hiding the entrance to a tunnel or cave that led to an opening on the back side to the, *"large stone open in the middle."* I was perplexed to learn the cascabel is a river containing small waterfalls. How many other rivers contained waterfalls? Also, Mejia said that the Rio Santa Rosa was the river to be explored, not a river named the cascabel.

"Adriano," I said excitedly, "we now know the Tambo Santa Rosa lies on the south bank less than 200 meters from Antonio's ranch and the Colombians named the Rio Cunquinza the Santa Rosa."

"Alfredo was right all along!" he admitted. He was feeling guilty for being so steadfast that we search the Santa Rosa River near the Ojal. We were both curious why there were two major rivers named the Santa Rosa including the little arroyo.

"Alfredo may have been right about the Cunquinza but he was wrong in the way he conducted himself with the Shuara," I said. "Necta wants nothing more to do with Alfredo. And I opt we terminate Alfredo's connection with the Shuara. Not his agreement with us on the project, only his direct involvement with the Shuara. But there's something else bothering Necta. I think he's hesitant about crossing the Tutanangosa/Tungus River."

I showed Adriano the notes I had made from the trip, and continued, "Here's the word I wrote down. I didn't understand his meaning to it. He said that beyond the Tutanangosa is 'represar'. I told him that I didn't understand. Then he used the word 'maldito.' Maldito means cursed.

"Did Necta mean that the south bank or what is referred to as the Kionade is cursed? I know there are many curses placed in different areas of this country. Guayas, the Indian chieftain placed a curse on Guayaquil prior to his death. When Atahualpa's captain of the Inca army, Ruminagui, learned of his emperors death, he placed a curse on the Llanganatis' and the treasure hidden there. Also, Quizquiz, another Inca general placed a curse on the city of Quito and the Spanish garrison that controlled the city. What do you think my friend?"

Adriano thought a moment before he answered. "In Spanish the word represar or embalsar means damming. With all the problems he's undergone there, the snake and tiger attacks and the awful weather; I believe he is making reference to that dammed place. But the other word maldito means curse in Spanish. I believe he is probably referring to that 'cursed damn place'. Although the Jivaro are very superstitious, I believe he is more frustrated than frightened with any curse."

Adriano was upbeat and pleased that the cable was on station in Sucua. The only thing bothering me was the timeline for the construction of the bridge over the Tungus. In this wonderful backwards country, it might be a week, month or another lifetime before the cable linked to that distant world. I also wondered whom Achacho and Necta had to seek permission from?

CHAPTER 14

DIFFICULT JOURNEY

Road into the Oriente

Weeks passed and I occupied my time studying Spanish and hunting local treasures in the city of Cuenca while I waited for the weather in the Oriente to calm. My interest in the hidden treasures of the community began when Adriano told me that his mother's chauffeur, a man named Pino, had found a previous occupant's treasure in his old house. It was so lucrative that he moved his family to the United States with the proceeds.

Then I heard about another find that stirred my interest. It was found near the Vintimilla's residence. Two workers had drilled holes in a residential building to anchor an RCA sign. They hit a major jackpot when their drill holes released hundreds of rare gold coins that peppered the street below and left traffic tied up for hours. The workers scrambled to retrieve the bounty. Unfortunately for them, the owner of the building had been alerted by the police and claimed the treasure was his because it was found on his property. A legal battle raged for months before a judge hearing the case ruled that both parties were to share in the discovery estimated well over half a million dollars.

The first electronic hunt with my metal detector began with a treasure dear to the hearts of the Vintimilla family. We were all aware of the friendly poltergeist that resided within the house and many felt it was trying to communicate a message. Mrs. Vintimilla was confident the discarnate was Miami Vasquez, the previous owner. It was rumored that Miami hid a treasure of immense wealth but died before he could reveal the exact locations. The rumors surrounding Vasquez were that he had spread his treasure between his finca (farm) and residence in Cuenca. With the country in perpetual revolution and military coups a regular occurrence, no one trusted the banks as the repository of wealth and instead chose to hide anything of value.

Adriano was sure the treasure resided somewhere in his house. Mrs. Vintimilla was thrilled with the idea of the treasure hunt, but her husband needed a little convincing. After a long discussion, he relented with a hint of trepidation.

We were upstairs in the master bedroom when my metal detector began squealing. Aha! We looked at each other, elated with our discovery, anticipating what we might uncover. We removed a layer of fine polished hardwood to find a large bolt with a thin metal wire attached.

"Maybe something is tethered to the wire?" I said.

Adriano agreed! We pulled up the bolt. Something of weight was suspended on the thin wire. We carefully raised the bolt higher. Suddenly the wired snapped and disappeared into the hole and at the same time we heard a thundering crash from the room below. We ran downstairs to Mr. Vintimilla's study, as did everyone else in the house. The patriarch of the Vintimilla clan sat at his massive desk with eyes the size of saucers, an ashen face and he shook like a leaf in a windstorm. He was surrounded by broken crystal from the chandelier that dropped from the ceiling and shattered into a thousand pieces. Seeing the mess we'd created we were relieved he was alive and sheepishly stood in front of him like two delinquent boys.

Adriano suggested I go to my room and he would try and explain our stupidity to his father. It was one of those moments I was glad I didn't understand rapid Spanish. For the next hour I cringed each time I heard my name mentioned in the room below. What a way to impress your hosts and not an auspicious beginning for hunting local treasure. I knew if there was a treasure in the house, it would never be found by my metal detector.

Word traveled fast in the city that the American friend of Dr. Vintimilla had a metal detector and that he would share any financial gains. Adriano and I scheduled a meeting with a family that, through a twist of fate, had hopes of great wealth.

A maid in a grand old hacienda was sweeping the floor of the spacious living room when she accidentally hit one of the tile panels on the wall. Somehow she had sprung a hidden latch, which revealed a small opening into a secret room. What was remarkable was the family that had lived there for seventeen years and never noticed the unaccounted space that housed the small room. Even more remarkable, the family was afraid to enter the small crawl space to see what might be hidden there.

When I arrived a crowd had already formed. Waiting patiently was the family, along with distant relatives and neighbors. After the exuberant greetings and handshakes my metal detector and I were beginning to feel like celebrities. They offered me beer, wine and other assorted drinks, which I declined and we gathered in the living room where the patriarch explained, with animated gestures, the challenge I was facing.

The opening of the chamber was wide enough to crawl through and I wondered why someone hadn't tried to enter. I asked the owner if he wanted to be the first to go inside? It was obvious from his reaction that he wanted no part in the initial exploration of his secret room. He waved me forward with a sweep of his hand. I crawled to the opening on my hands and knees and pushed the equipment through then squeezed through myself all the while wondering why the family was frightened?

I went headfirst into the dark chamber. I felt dirt on my hands and breathed in stale air that flooded my nostrils. I switched on my flashlight and saw that the room appeared to be about ten feet wide and eight feet in length. It was empty. In the center was a square, stone, alter-looking object. There was nothing on it or in it and I wondered if at one time the room had been the storage place of valuables kept away from prying eyes of outsiders.

I poked my head out of the opening and asked for someone to join me to hold the flashlight so that I could sweep the room with the detector. There were no volunteers. I shrugged, retreated into the musty space, placed the flashlight on the pedestal and began my search. Nothing pegged on the metal detector and there was nothing visible so I crawled back into the living room.

The crowd stood like statues. I shook my head and said, "Nada, nothing". Their hopes dashed, the group quickly dispersed without even a thank you. I packed up my gear in the presence of a maid who stood watch and unceremoniously left.

My next project had both prospects and pitfalls. In the distant past a wooden box the size of a coffin was supposedly buried behind a residence that had been demolished. Nothing had replaced the structure and it was now a vacant lot adjacent to a government office. Adriano researched the titleholder of the property and learned that the Banco Central owned it. We agreed that we'd conduct our own clandestine exploratory mission and if we got a reading he would obtain the necessary permissions.

Under cover of darkness and with Adriano as my lookout, I searched the location with my trusty detector. It produced a positive reading that matched the description and size of the treasure buried on the location. I estimated the reading indicated an object just over two feet wide and seven feet long. We were elated! Adriano made a call and arranged a meeting with the vice-president in charge of property management for the bank.

An old chum, he greeted Adriano and we quickly brought him up to speed on our first foray and then made our request. He got up from his desk, placed an index finger over his mouth and approached the door to his office. He poked his head into the outer office, and then closed the door.

He asked quietly, "What could be buried in the coffin?"

Adriano explained that if this was the rumored lost treasure it might contain anything from gold and silver to priceless jewelry and maybe the legendary green book that held the maps and locations of the Church treasures. When the book was mentioned the banker's eyes widened. After he regained his composure he said that a meeting with the board of directors might prove non-productive and he preferred to keep the venture confidential. If we would share the find with him then he would take personal responsibility and allow us to dig on the property. We agreed. I explained that the other problem was that the location was too open to curious onlookers and that a fence would need to be built around the property to insure privacy and security during the dig. He promised to have a fence constructed as soon as possible so that we could begin an earnest search of the area.

Just when local treasure hunting was getting interesting, Necta arrived in Cuenca unannounced. With him was Antonio Ambusha, a Shuara friend and he had good news. Ambusha was willing to help us locate a waterfall that he knew as a youngster as the cascabel. Their arrival and news sent me reeling into another expedition mode. In the ensuing days, Adriano and I bought supplies, checked our maps and located a land rover with a driver. Only one glaring problem remained. The waterfall was on the south side of the Rio Tungus and we must find a way to cross it.

We left Cuenca before sunrise in a Land Rover whose seams swelled with men and supplies. The first challenge would be the journey itself. Through luck or misfortune our Rover would be the first vehicle to make the six-hour, ball-buster journey on the new road that took years to carve out of the mountains and jungle. My passenger list included Antonio Necta, Ambusha and a driver that seemed to have lockjaw. The money was too good for him to pass up and I couldn't tell what worried him the most, the road or the Shuara sitting behind him staring at the back of his head.

I checked my army map. The road we were on didn't exist. The only rivers listed were big rivers and they were few and far between. My circa 1950 map, the result of data provided by the Geographic Institute of the Ecuadorian Military, The U.S. Army Map Service, I.A.G.S, and Shell Oil Company showed the Tungus forming in the Cordillera Ayapungo at an elevation of 3500 meters. This too, was worthless. My only true compass was Necta's map, a crude drawing on stained paper that started in Sucua, went upstream along the Tungus and listed the names of nine rivers that required crossing before reaching the Rio Ojal. The Ojal or "buttonhole" ran from north to south at an elevation of 2500 meters. Beyond that there were no rivers, mountains, game trails, playas or villages noted. It was blank space with a river named the Tungus running through it.

Once we reached the serpentine artery it would be our navigational beam. Our plan was to follow it west until we found a shallow area that provided a crossing point. Hopefully, we wouldn't be forced to climb all the way to its source with full backpacks in rarified air. It was heartbreaking to know that a cable bridge across the Tungus would have saved a week of a backbreaking journey.

But first we had to traverse an invisible road that became an exercise in patience and fortitude. On the passenger's side steep cliffs descended into a far off river below while on the driver's side of the vehicle large rocks balanced precariously overhead. Vigilance was the by word when the fog rolled in and our field of vision narrowed to zero. When the fog lifted the rain came and with it the mud. The brown ooze eventually took control of the steering and sporadic traction caused my stomach to churn in sync with the spinning tires.

After an hour of touch and go travel we reached a concrete bridge suspended over a chasm that separated us from a churning river below. Necta told us that we were above the Rio Negro. This was the bridge that took years to build and was the final piece of the road that linked Cuenca to the Oriente. This bottleneck was why I had made so many DC-3 and charter flights. I had waited a long time for its construction.

The harshness of the terrain was compounded by a deep chasm where two turbulent rivers, the Negro and the Paute fought each other for dominance. The concrete and steel pylons that supported the bridge constantly took a beating from the crashing swirling water. Hopefully they used steel for reinforcement, but in this country, quien sabe—who knows?

Our driver in need of relief stopped the Land Rover in the middle of the bridge much to my dismay. We departed the idling vehicle and gathered above the ambiance of the boiling rivers and marveled at the fact we were able to stand upright after hours of sitting. I stared at the road we had just come down and couldn't help but notice a huge mound shape hill sitting like a majestic sentinel overlooking the rivers. Necta noticed my interest and said the name of the mountain was Chupianza, which meant "place of gold". He didn't know why it was named that, but it was. There was something compelling, something strange about this place. I knew that one day I would know Chupianza and it secrets.

Once we crossed the bridge the terrain was dotted with patches of banana trees that blended in with a few thatched huts from locals that farmed the area. In the valley below Necta pointed out the small military garrison in the village of Mendez. A few hundred yards above was the towering rainforest in columns of different shades of green, some with silver tipped leaves. Small streams seen in the distance looked like white ribbons fluttering in the wind. They disappeared beneath wooden plank bridges that were nothing more than a couple of two by fours hastily placed across the streams.

Our emergency stops were frequent. When we approached one of the many small bridges constructed from wooden planks lying side by side we'd depart our vehicle and check their strength and durability. Sometime the boards were separated and it would take all of us to put them back in place. Then I would slowly guide the driver across the makeshift structure; it was like positioning a vehicle on a lube rack. I couldn't imagine a bus trying to cross these planked bridges much less driving the road at night.

It was impossible to cushion the pounding on my butt and legs from the rut-ridden road and despite a dozen or so seating readjustments my efforts only offered temporary relief. Mental escape came through my constant stare at the ever-changing scenery and bird life. We rounded a curve where the road arched into a semi-circle of flat, empty terrain and a sudden movement broke my catatonic gaze. A crimson red snake about seven or eight feet in length with head high zigzagged in pace with our Rover. The red rover was outrunning our Range Rover.

I rolled down my window to get a better look at the magnificent reptile to the dismay of the driver, who also noticed the snake had become curious with us.

"Please! No, Senor Lee," he screamed in fear.

About that time both the Antonio's in the back seat let out an, "Ajiie!"

They started screaming their name for the snake in unison. "Dos Minutos, Dos Minutos."

This was a snake that persecuted a man and the only recourse was to stand and fight because a man couldn't out run it. I understood what Necta and Crespi had spoken of when they said that to survive an aggressive snake, a man must stand his ground with a machete and wait for the serpent to enter the kill zone.

Eventually the Dos Minutos lost interest and veered into a patch of jungle. We entered a flat stretch of land and passed a side road that indicated something of importance was out there. Necta noticed my curiosity and related a morbid story about the dark side of the Jivaro struggle for justice against the Spaniards who invaded their land and lulled the proud tribe into a false sense of security for the sake of gold. The time was 1599 when the Spaniards were combing the country in their endless search for the tears of the sun and had forged an uneasy alliance with the Shuara of the

Upano Valley.

Beyond the mountains was—at one time—the Spanish village of Logrono. Back then it was a colonial city founded for the purpose of working rich gold placer deposits and it was there the Spaniards developed gold fever. After the Spaniards subjugated the Indians they required a perpetual tribute of gold. As the years passed they never seemed satisfied with the offering no matter how much of the yellow material was given to them.

Finally, the Shuara rebelled through the leadership of chief, Quirruba and decided enough was enough. Spanish inhabitants died by the thousands in their homes. They killed everyone except the governor who was saved for last. He was led into the square in his sleepwear and told that it was time for him to receive his tax of gold. They pried open his mouth with a bone then poured hot molten gold down his throat until his bowels burst. Antonio concluded his story with, "Lee, his appetite for gold was finally satisfied."

I laughed along with the Shuara while our Spanish driver slumped over the steering wheel, stared straight ahead and prayed that we would reach Sucua before long. He shot me the frightened glance of a man without a friend among a group of heathens. He was undoubtedly questioning my sanity. I knew he had to be asking himself why an American would take up with a group of rebellious Indians that had no respect for authority and then join in the camaraderie against his civilized culture. His fear played well with my savages.

Once we reached Sucua our driver made a hasty retreat back to civilized comforts. We stretched our legs and later bedded down in Necta's small house. The first phase of our long journey was complete; we had survived the drive on the new road. After a night at Necta's Sucua home and another at his finca Wakani, Necta's friends, Macas and Gonzalez, joined us and we began the second, more tasking phase of the journey that took us west along the Tungus in search of shallow water to cross into the Kionade.

Luck was with us on the second day when we found a hunter's path through the high jungle and reached the Rios Tunanza and the Conquinza. We made good time and reached the east bank of the Rio Victor. Black clouds loomed ominously to the west, a preview of where we were headed. Necta was visibly concerned. He knew the Tungus was taking the brunt of the runoff and this meant that the narrow points of the river would be widening and we would be hard pressed to find a crossing point. To the north the sky was cloudy, but not foreboding, which meant the rivers to be crossed had not yet grown angry and aggressive.

We picked up our pace as the sun danced through the rope-like tangle of undergrowth that embraced the trees that formed a semi-canopy above. Our rest breaks were always in small clearings that allowed us to loosen the packs, stretch, and soak up the sun's rays. We, like the plant life and reptiles, found ourselves constantly in search of the sun's warmth and infusion of life. There was nothing more ominous than plowing through a dark jungle. It was similar to a leafy, tangled tunnel with a terrace above that filtered out radiant heat and light. In the rain forest, semi-darkness and decay created a stranglehold on the spirit and eroded confidence with gloom and despair. For me the passageway of vegetation required mental adjustment, otherwise despondency ruled.

When I passed the beef jerky around I couldn't help but notice Ambusha's fascination with his newly acquired AR-7 and its lethal magnum hollow point ammo. His excitement was real. Now he had a semi-automatic with real bullets. He acknowledged my gift and caressed its stock and barrel with reverence, a befitting gesture for a man that previously entered the Kionade with a single shot 22 and with what I referred to as target practice ammo.

At first he thought it was just his to carry on the journey and it took several minutes for Necta to explain in their native language that the weapon was not a loaner, but his to keep in exchange for services rendered. When he realized this wasn't a natema-laced dream, he couldn't thank me enough. He didn't know how. That was fine, but the look he gave me sent shivers up my spine. I knew

I had my own Kakaram, a powerful one, a killer with a magical weapon and bullets that would dispose of anyone or anything in my displeasure.

For a Shuara to be accepted as a Kakaram, he must have killed at least a few people—whether the victims were from inter-tribal raids or for the coveted tsantsa was unimportant to the killer. They, like missionaries, collected souls and passionately believed in their quest. Ambusha easily fit into that mold now that he owned a weapon of accuracy and firepower that was greater than his rivals. However, now that Necta had taken him under his wing, I felt the age-old Shuara tradition was no longer of interest to him. His respect for Necta had changed his interest from death to life.

The AR-7 gave him an edge in the game of survival in the unexplored places he was drawn to. This was the first time a person had given him something that was not broken or discarded. For this he was appreciative in his own silent and guarded way. It took some prodding by Necta before he approached me, nervous and embarrassed, he shuffled his feet like a kid in a first grade recital. He avoided eye contact and displayed his first emotional response. He mumbled what I thought was a quick thanks.

It was getting dark when we arrived at the Rio Huahuayme commonly known on the colonial maps as the Tigress. We pitched camp on the east bank among leafy trees and tried to relax, aware that at first light we would be ropedancers fording the dirty brown river on slippery logs and rocks. Rotten Acapu logs fueled our fire and gave warmth while we drank manioc beer and ate saltines topped with tuna, sardines, rice and beans. My plan was to eat as many of the canned goods as possible, thus reducing the weight in our backpacks before we began the constant climb.

Before the remaining sun turned shadows into darkness, we placed our tents in a semi-circle with the river at our backs giving us one less direction to defend. For the first time in months I felt free and unlike the past when I couldn't wait to get out of the jungle, I was happy to be there although we were knocking on the door of the forest guardians and the niawa (jaguar) packs. There was an anxious feeling to know that in the morning we would enter the private hunting ground of the predators that didn't take kindly to outsiders.

The next day we successfully crossed the Huahuayme and the Curinza rivers without any major mishaps. Our pace had quickened to the point we were bearing down on the first legendary Santa Rosa River named by the Colombians after the first saint of the New World and the patroness of South America. It was my first look at the mystical river that previously had us convinced that it was where Mejia found the emeralds.

When we stopped to allow ourselves a deserved breather, I heard the two Antonio's conversing in their native sing-song-dialect. I would have given anything to be able to join in and converse with them in what sounded like a simple but beautiful language. After the conversation, Necta translated what Ambusha said. According to an old Indian that used to go hunting with him, the Santa Rosa River was originally named the Rio Imuirtni.

Ambusha's education was limited at best and when I asked him to write the name in my notebook, he hesitated. The name given to me was done so phonetically and the spelling was open for debate. It was similar to the word Kionade. I wound up with three different spellings for the place. The mysterious place had been labeled as Kyonaide, Kionade and Kaionede. They all knew the pronunciation, but for them to spell a name was a gamble at best. So, for my purposes, I settled on Kionade as the mystical name of our destination.

We camped alongside the Imuirtni, the second Santa Rosa River, and the area explored in detail by Bill Wright's expedition. Unlike then we were still in good shape weather-wise and although a storm was gathering force to the west, we were dry. We huddled around the campfire and told stories about our lives and why we enjoyed the primitive lifestyle. I began by telling them about growing up on the Apache Indian reservation in Arizona when life was simple and hardships were common. When I told them that the Apaches reminded me of their tribe, with the exception that they only took part of a man's head, his scalp, rather than the whole head they looked bewildered as to why they stopped before taking the best part. Ambusha in particular failed to understand why the scalp was more important than the enemy's soul.

After a lengthy conversation, I was asked why the colonials defeated the Apaches if they were so fierce and so brave. I responded that the Indians were outnumbered and outgunned and, like other great tribes that faced the white men, they were seduced into believing they could eventually live in harmony together. They entered into treaties that became a trail of tears, broken promises and deployment to reservations where they could be monitored and governed until they became "civilized". Unfortunately my history lesson fueled leftover anger for the colonials that had stampeded into the Upano Valley in search of farming land, timber and gold. It was best to change the subject.

Each man was asked, with the exception of Necta, why he enjoyed the adventure of uncharted territory. Gonzalez responded that as a youngster he had heard exciting stories from his father and this was his opportunity to experience it first hand. He, like his father was interested in nature's abundance beyond the Ojal. Macas' answer was that he craved freedom of movement and expression from those he didn't trust and a civilization that regimented his lifestyle. Ambusha remained silent and only occasionally nodded in agreement. Then I asked the withdrawn one the same question.

His answer was more complex than I was prepared for. This quiet and reserved loner was slowly leaving his shell. He responded with self-confidence and bravado that this was not unexplored territory. He said the first Kionade would seem unexplored even if we were there for months. It was a vast land inhabited by supernatural servants, some good, and some bad. The magicians lived there. He said I wouldn't understand why I must go there until I was there. Was he a sage? I hoped he was referring to beauty rather than danger because if he was a danger junkie then we were in for one hell of a ride.

That evening as the fire burned low and the men strayed into their own dream state, I found myself unable to sleep and rehashed Ambusha's cryptic answer of the lost world just across the Tungus. What did he mean by first Kionade? Damn, I thought to myself, does everything up here have a twin?

The next morning it took the better part of an hour before we crossed the turbulent Imuirtni chained together with fiber from the Cabuya tree that was stronger than rope. We were soaked and our packs drenched before we reached the other side as we slipped and slid into and out of the white foamed river. On my feet were the best hiking boots money could buy and they were worthless compared to the cheap rubber irrigation style footwear of my pack. When their boots filled with silt, sand and water they just yanked them off their feet and poured out the contents. When mine filled, I had to stop and untie the laces then pour out the water. Needless to say water and leather were strangers to each other. We didn't bother to dry ourselves and pushed off into the beginning of a tense day.

There was a nervous stir among my team. The thickness of the jungle left us defenseless as we climbed and crawled on all fours through a green maze of tangled growth. The pain was unbearable. The harness straps from my backpack cut into my shoulder blades and caused me to wince with each shift of weight. My lungs struggled with each gulp of air and my wet clothes clung to me like a thorny blackberry bush.

We were in single file with Ambusha up front in the point position followed by Antonio and Gonzalez, then me and our human mule, Macas. He was carrying his and Ambusha's load so that our point man could be free to maneuver with his weapon in case of any surprise. Even with his extra weight he kept bumping into me and then would back off and give me space. Bill's prophetic statement came to mind. "Hoss, I never thought I would say this, but the area up there is too much for a white man."

It was beyond difficult; my body was on the verge of surrender. We climbed one incline and slid down the next until the mind went numb and I lost count of the grueling cycle. When we finally reached level ground I was temporarily relieved until I saw the massive field of cana brava that stretched before us. The wild field of palm cane used for ceilings on houses swayed lazily in the afternoon breeze. The serene scene was misleading; before us were razor sharp fronds higher than a man's head that offered little comfort or safety. But we had no choice. It was scary to enter blindly with limited senses dulled by the deafening sound of the Tungus and cracking thunder but we ploughed into the thicket.

There's rarely a moment in a palm cane thicket that one feels comfortable. The constant movement of the swaying plant life played havoc with my nerves. When Ambusha unslung his rifle, I did likewise knowing full well that if we accidentally stumbled onto a predator we would have little time to react. The element of surprise would be to our advantage, unless we interrupted the Dos Minutos mating ritual or a cat feeding off its kill, then of course we would be in a world of hurt. Finally we cleared the thicket and all took a deep breath.

Reaching the Ojal was an arduous excursion. The river named after a "buttonhole" lived up to its name. Towering trees formed a giant canopy of greenery that surrounded us. It looked like a monstrous spider web that obscured the sky above. Without warning the "tubo", a tunnel cut into thick overgrowth opened onto the river strewn with large boulders and swirling pools of white water. The Ojal wasn't very wide, maybe forty feet, but it was deep and swift. It was a formidable obstacle to cross unless we could fell a large tree for a makeshift bridge. If we had a power saw, no problem. But all we had were four machetes to chop a tree that was long and strong enough for men to crawl across. We were blocked from moving any further and as if summoned by the spirits, it was beginning to rain, light at first, but soon torrential cold drops the size of a sucre coin pelted us.

Gonzalez and Macas cut out a campsite while Necta and I went up stream to search for a narrow opening to cross the river. Huge Guayacan trees blocked our path and proved impossible to fell.

Guayacans are known throughout the Amazon to be the hardest wood and, although difficult to carve, were used by the Shuara to fashion their lances for hunting and warfare. Thoroughly soaked we gave up and returned to the campsite to regroup and rethink our next move.

It was difficult to hear each other over the roaring Tungus to our left and the crashing Ojal in front. We screamed at each other as we struggled to hoist the tents in the wind and rain. Once inside we munched on beef jerky to regain strength and huddled together to stay warm. The Shuara, no strangers to nature's fury, were laughing and making the best of our situation while I shivered and shook like a leaf in a hurricane. Ambusha reacted to my condition, reached into his backpack and pulled out an old discolored sweater and placed it around my shoulders. It was the Shuara way of offering something in return and I readily accepted

In a steady rain we dug a trench around the tents to drain off standing water while Necta cut a large stick and pounded it into the bank of the Ojal. This would be our water marker and tell us at first light how much the river was rising or receding. I checked my watch; it was a few minutes after four p.m. Still, I fell asleep in the corner of the tent and drifted in and out of my stupor. It was a long night because I really wasn't sleeping. I was just waiting. Waiting for the rain to subside and waiting for a surprise, just waiting while feigning sleep. Subconsciously I was praying our tents would hold from the pounding they were taking.

"Shades of Little Hell," I told myself before I finally drifted off.

The next morning our confidence wilted. The water marker had been washed away during the night. It was time to cut our losses and get the hell out while we could. In a steady rain our tents were struck and we headed due east in silence. Necta was dejected; once again he felt failure—whether from nature's fury or a wawek curse—it really didn't mater. In his mind we had failed at another attempt to cross the Tungus and it gnawed away at him. The Kionade was still off limits and would be so until the Tungus ran shallow. For me it was an acceptable fate because for the first time in my life, I wasn't pushing to make things happen at the expense of my emotions and body. My agonizing experiences in this wild country had taught that the more you push, the more *IT* pushed back.

CHAPTER 15

WAKANI

Wakani

It felt good to be back at Wakani, the spiritual home of Antonio Necta and the gateway into the unknown. The timeless Tungus, loud and defiant rolled eastward in search of the mighty Amazon River. It had rained incessantly the previous week, the ground was still muddy and the place smelled of mildew. Today the sun shown brightly as we laid out the backpacks and assorted supplies to dry and clean.

My days at Wakani were filled exploring the Tungus with Necta's batea. Ambusha returned to his hut along the river to the east while Gonzalez and Macas took horses to Sucua to replenish our canned goods and spend time with their families. The waterlogged terrain slowly began to return to life from the warmth of the sun as evidenced by new growth sprouting in Necta's garden.

My evenings were spent in private language school. I taught Necta new words in English, he likewise schooled me in Shuara. A machete was machitrumin. A plant was another tongue twister: arakmariin. What is your name was ame naaram yait. I loved the word, why. It was urukamtai. While I tried desperately to at least learn a few words, my counterpart was having his problems as well. That made me feel a little better and we laughed at each other's poor pronunciations.

Although the mountain-jungle chickens were tougher than a boot, the mouth-watering vegetables, fresh from the garden, uncontaminated by pesticides or artificial flavoring certainly made up for the jaw breaking fowl. After dinner it was story-telling time around the open fire with me prodding Necta to tell his adventure stories. After a couple of beers he would oblige and with eyes growing wide and his voice reaching another decibel he would share tales about hunting trips into the forest.

I was sure he had told the stories more than a few times; yet, he imparted the excitement as he told of the time a tigre attack almost took his life. He was in his teens hunting with his bodecaro (blow gun) when a medium-sized cat attacked from a ledge above him. He said his quick reaction escaped the brunt force and the cat only grazed him. Although wobbly he managed to place a lethal curare dipped dart into the blowgun's chamber as the jaguar regrouped for another charge. He propelled the poisonous dart full force into its cheek. The animal began pawing in a desperate effort to remove the stinging object of death. But it was too late; the brave jaguar let out a grunting sound and dropped dead.

When he finished the story that was accompanied by animated movements that mimicked his escape from the feline, I felt relieved he had made it, even though the outcome was obvious. Over time he had killed numerous jaguars, some small and many large, with his firearms, but that was his first kill with a bodecaro and elevated his status among his peers.

With softness in his voice he said that years ago he learned from within that to kill these beautiful creatures for their hides was self-serving and an insult to nature. He scolded those that refused to live in harmony with wildlife as the animals too, had the right to survive and when he saw pelts being sold for profit he said his soul wept for them.

Necta was unlike many of his tribesman that hunted for sport. I was beginning to understand why the Shuara zealously protected their land. It wasn't just the rivers that would suffer from being poisoned with mercury if large amounts of gold were found in them. I realized that discovery of the emeralds would bring in the bottom-feeders who would rape the land for the sake of profit and greed. The lives of these people would change for the worst, nature would be raped, and the wildlife would be exterminated for their skins.

I was beginning to see the light at the end of my dark tunnel and it was time for some soul searching. Did I really want to continue my search for Mejia's emeralds? Did I want the responsibility of bringing greed to this amazing part of the world? It was a mind-numbing decision process and painful to make. Nevertheless, I was leaning toward abandoning my search for the green stones. I made the decision to spend another couple weeks at Wakani, weigh all sides, and then return to Cuenca.

One mystery still remained to be discovered. One evening after his family returned to Sucua and we were alone drinking beer and swapping stories, I had the opportunity to ask about the snake experience with the forest guardian. He didn't hesitate and seemed comfortable that I would now understand the significance of the event. Back then, it would have been just another rain forest myth and open for debate.

He filled our cups with warm manioc and true to Shuara protocol began a lengthy series of events before answering my question. We sat on the floor as he spoke in a slow and meticulous manner to make sure I understood the incident preceding my first trip to Wakani. Then he blindsided me with the question we hadn't had time to pursue at the Vintimilla residence during our first meeting. "Lee, do you believe in spirits."

All that was required was a simple yes or no response, but I wanted to ask a question of them, "How do you define a spirit?" I knew that answering a question with a question was an awful offense to the Shuara, so I decided to hold my question for another time, nod an affirmative, and wait for his response.

He replied, "It's good that you believe in the spirits because they are the ones in control of our destiny in the Macas Forest."

When he accepted the responsibility to become our point man to explore the area that we believed contained the Mejia treasure it was done through friendship and trust. But, during that expedition he believed his efforts were being undermined by a wawek (bad spirit) who influenced snakes and tigers to attack and curtail his efforts. He was concerned, but before he could find answers he felt it was his duty to accompany Bill Wright on his journey up the Tungus. It was during that journey he discovered the expedition was accompanied by greed for the green stones.

He paused for a moment before discussing why the expedition failed. Then he softly repeated the word avaricia, the Spanish word for greed. He said that he and his Shuara brothers were not seduced by the greed of the Mejia treasure. Their intentions were motivated by survival and making a living to support themselves and their families and nothing more. But greed did accompany that expedition and was always arms length from him the entire journey. He had seen the greed in the eyes and actions of Alfredo and noticed that Bill was easily influenced by the lure of the green stones and by the whispered talk of Alfredo. He questioned that if Bill was that easily influenced by Alfredo, was Doctor Vintimilla and I also under his spell?

Necta said that in the beginning his search for the truth took him on a journey to the Village of the Shaman, a place deep within the rainforest where uwisins went to gain knowledge and power from the panu. With a little chuckle he confided that when he asked for my Polaroid camera months ago he had developed a plan. I asked what he did with the camera that had delighted his wife and he launched into a description of the whole adventure.

It was a long trek to the village where the Canelos tribe lived and it was late afternoon when Antonio changed into his native attire that resembled a kilt held together by a bark string. He replaced his baseball cap with the traditional estemat, a woven cotton band with toucan feathers, shed his denim shirt and carefully placed them in his backpack. Although the Canelos tribe had been in contact with miners, missionaries, and traders, Antonio would enter the Canelos village of the shaman as a proud uwisin of the Untsuri Shuara tribe and was determined to be welcomed as such.

The Canelos shamans were distinguished with a special term, panu, derived from banco, the Spanish word for the "bank". Canelos shamans call themselves "banks" because they held the wealth of magical knowledge and power available in the Oriente and were sought after by the shamans of other tribes.

Antonio was after the special magic that only the Canelos possessed. It was named the "white man's tsentsak". He knew that his expeditions up the Tutanangosa had been compromised by a powerful magic that had directed the cats, snakes and weather to prevent his progress. He thought it had to be the white man's magic that was controlled by an evil wawek tribal member. That was the only thing that made sense to him.

Canelos shaman had been successful in capturing and possessing the white man magic through years of study while trading with them. Antonio had heard the rumor that the possession of the magic took place when the colonials or white men were killed and their heads severed from their bodies and the mouths and eyes sewn shut to capture the soul. This timeless tradition among the headhunters of the Amazon gave the power of the captured soul to the warrior who made the kill and, if he so happened to be a panu, which was the name given to the Canelos shaman, then his power increased ten-fold. The white man tsentsak was more powerful than the Indian tsentsak because of the white man's daring and expert manipulation of people and truth.

The most powerful magical darts and the most valued magical darts belonged to those with colored eyes and forked tongues; those that traded steel axes and muskets that spoke death from a distance for the yellow pepitas. The Canelos panu discovered the white man had more power than the Indians, because those who spoke the Spanish would be protected by their powerful military and their priests in black robes that spoke of one God, in the name of God. The Canelos, who spoke in a Quichua dialect, had prophesied that the Jivaro days were numbered and that the white man would soon own or reside in all of the lowlands surrounding their territory.

Because of this, the Canelos objective was to capture and hold this power, which they did. The panu possessed the powerful tsentsak of the colonials and were the only shamans in the region who were able to become possessed with the souls of the dead, which enabled them to act as oral mediums. They had the power to send demons to possess victims and manipulate their behavior. They also had the cures for the "white man" diseases, such as leprosy, whooping cough, measles and gonorrhea. They had become so powerful they believed they could not be killed by ordinary means and this was the opiate that attracted lesser shaman into their midst.

The Kakaram also came to seek the white man tsentsak, which gave them more power to carry out their roles as killers. These outstanding killers were the desperados of various tribes and distributed their own brand of justice for the people of the tribe who requested vengeance in the constant raging inter-tribal wars. There were only three socially accepted and prestigious titles in the Jivaro world and evil reigned in two of the three. There were pener uwisins who cured. The wawek were the uwisins who bewitched their victims and sometimes used magic to kill. And last, the Kakaram, who solely killed. Sadly, Necta remembered that these were the ones the young men of the tribe wanted most to be, but in order for them to become an outstanding killer they had to acquire the arutam soul power that could only be achieved by killing several people.

When they started this life the young men went on 'honorable killing expeditions' to avenge the deaths of their fathers and other close relatives to achieve the arutam soul. That honorable quest turned into an ominous future for many of them. Their reputations would follow them as the best in their field and create a life and death struggle for them to remain on top. Like an old gunfighters from the west, the new Kakaram constantly challenged them. They were men who wanted the reputation of the one who killed an outstanding killer. These were the ones that Necta feared most because their boundaries of decency were blurred by the lust for power through death.

Necta entered the Village of the Shaman at dusk and followed the dirt trail between interlocking roundhouses that straddled a small tributary named Tunura, the Jivaro synonym for Hell. He made his way through the maze of houses and people careful to avoid direct eye contact with the many different cultures that were also on vision quests and trading missions. Out of the corner of his eye he recognized the Kakaram by their lifeless eyes, shotguns, and rifles. Occasionally he would see an outstanding killer whose prestige was augmented by a 44 Winchester strapped over his shoulder; the better the firepower, the better the reputation. He felt sadness for these young warriors who had come to barter their souls for additional soul power; all they had to offer a panu was their own soul.

As he turned a corner he accidentally bumped into a tall, muscular Achuara warrior who instantly recognized his tribal markings and then looked him straight in the eye. Necta dared not flinch or look away; this would be a sign of weakness and fear. Necta returned the steady gaze and then both men stepped aside and continued on because in the village of the shaman there was an unwritten law of neutrality.

The Village of the Shaman was a safe haven for those in search of the forbidden knowledge and no one dared to violate this edict for fear of facing the wrath of the panu. The shaman knew the story of the two warriors from combative tribes that waged war against each other here and the one that survived the deadly encounter died a horrible death that same evening while he was sleeping. It was rumored that a deadly Coralito snake had been sent to him while he slept. Needless to say, the story circulated like a distant drumbeat throughout the region. The law was cast: if you take blood in the shaman sanctuary, then you must give blood and your life.

Dusk along the bank of the Rio Tunura brought the isolated compound to life. Fires were lit and torches strung along the crude storefronts on the merchandising side of the river. Passing through the tangle of humanity that was eating, laughing, singing and shopping, Necta stopped to admire the various skins of wild animals being bartered for, lying lifeless along side captured animals in cages howling in fear sensing their imminent death. Those untouched by sensitivity of nature laughed and made fun of the howling creatures begging for their lives. Man has to be the cruelest of all of nature's creatures, he told himself.

The Achuara trading bins were stuffed with blowguns, curare and native beadwork they hoped to trade for steel cutting tools and guns. The frontier Jivaro residing west of the Cutucu Mountain range were in direct contact with the whites at Macas and Sucua and displayed colorful colonial clothes alongside portable transistor radios next to Canelos bins displaying monkey, jaguar, ocelot and soft peccary hides, 44 magnum cartridges, and beautiful gold nuggets that were fashioned into crude jewelry.

One trading bin had rows and rows of shrunken heads that seemed to range in shape and size from young to very old. Although the ear lobes were adorned in colorful red and white toucan breast feathers, they all had the familiar grotesque eye and lips sewn shut. Necta said he felt ashamed as he gazed at the tsantsa of forgotten human beings dangling before him.

There were moneychangers, eating and drinking establishments and a large building that served as a sleeping area. He crossed the wooden bridge suspended by rope vine that separated the two worlds of commerce and magic and remembered the instructions that other uwisin had given him. He was to cross the bridge and look for the largest roundhouse in the area. He was to seek out Kankiam, the ancient panu who held all the power and knowledge.

As Necta cautiously made his way across the unsteady and swinging rope bridge he noticed a change in his demeanor. He felt as if all life had been sucked from him, subdued by an unseen presence that seemed to be monitoring and controlling him. There was no joy, happiness or fear on this side of the river. It was as though all human emotions drained away making him a pawn for the chess masters of human spirit that awaited him in the void of semi-darkness.

There were many roundhouses on that side of the river and in each lived a panu with special powers and magic. The larger the roundhouse meant the more Kakarma, or power and prestige, the panu wielded. Many were evil and dealt in death and destruction. The Kakaram and the wawek sought their knowledge. And there were those panu who supplied the medicinal needs of the pener uwisin who sought the newest forms of cures for various illnesses. But, there was only one panu who held all the knowledge and he was the most feared and most sought after by all that made the perilous journey into the dark side of the Amazon. He lived in the largest roundhouse nestled next to the Tunura and Necta knew that he would only have one chance to gain access to the legendary panu named Kankiam.

Approaching the largest of the roundhouses Antonio was startled by a voice that came from the shadows and asked for his name, tribe and intentions. Caught off guard Necta managed to answer in rapid fashion, quickly adding that he was a friend of Achacho, a respected, old pener uwisin from the Untsuri Shuara. Silence greeted Necta's response. It seemed like an eternity before he was asked to enter the private world of the Canelos panu.

Accompanied by a young warrior who introduced himself as the grandson of the man he had come to see, Necta entered the house and noticed that the single gabled roof was made of closely woven kampanaka palm thatch. Countless center posts cut from the Cedro trees supported the large structure. Each post had a sacred animal carved into it, which added to the mystique of the residence and those who inhabited it. Windowless walls were constructed from the chonta palm and were arranged in a way to allow air and light to enter which gave the dwelling a feeling of freshness and freedom. Posts, beams, roof and walls were all lashed firmly together with tree bark and the floor was covered with fine sand from the Tunura. The house contained two separate areas, two halves: the men's and the women's. The man's half also served as a parlor and sleeping area for special visitors.

A beautiful young girl placed a tray of manioc beer and a clay bowl containing a brown root residue that resembled the sacred natema on the floor before him. He was asked to remove his sandals and empty the backpack on the floor and wait for further instructions. Then both the young man and girl left. Somewhat apprehensive he sat down in the sand with his eyes closed, legs crossed, hands with fingers placed upright on his thighs, and waited for instructions.

Within minutes a strange energy shift alerted Necta that someone was in the room. He opened his eyes and saw seated across from him an old man, dressed in a Canelos kilt, sandals, and colorful toucan feathers that accented his shoulder length gray hair. A beautiful stone, deep green in color dangled from each ear lobe. In his right hand Antonio noticed a pure white quartz crystal that reflected the orange rays of the burning torch in the background that added to the mystery of the moment. Antonio was poised to speak when suddenly a cooling sensation crossed his forehead somewhat like a gentle breeze and a thought voice told him to relax and have some beer and speak with truth concerning his purpose for being in the old soul's presence.

Reaching for the earthen pot of delicious manioc beer, Necta carefully structured his thoughts while framing his first request. He had to be careful because he knew that his thoughts were no longer private, but this did not bother him as he knew that his thoughts were pure and without vengeance. Without hesitation he poured himself a drink that was cooled from the running waters of the Tumura. It refreshed his parched throat and dry tongue. He was thinking that it was the most delicious beer he had ever tasted when the silent voice responded, "Yes, I know."

Kankiam had displayed one of his powers and was fully defining the boundaries to his guest. Then he introduced himself in a deep, but gentle, audible, voice, which made the uwisin from Sucua more comfortable in his surroundings. Kankiam, who obviously came to power because of his intuitive abilities, was curious as to why a proud pener uwisin who possessed strong Kakarma would want the deadly and powerful white man tsentsak, especially since Necta did not have dishonorable intentions towards his half brothers.

"You mimic their dress, speech, religion and worldly goods. You are comfortable with this yet you want this magic. Why?" He asked.

Necta knew that Kankiam had the answer to the question before he asked and was testing his integrity and honesty.

The manioc fueled his tongue and Necta poured out his frustrations of being a pener uwisin caught up in two worlds with two different ideologies from opposing forces. He explained the violent and aggressive attacks from the tigres and the whistling snakes and the feeling of hopelessness that he disguised from his band of brothers. He knew that the attacks were directed at him and his expedition by a force clearly beyond a wawek power and he surmised that this magic was from a "white" source. But, he had no proof and he had journeyed this long distance to seek the answers to his questions. He needed the white man tsentsak to balance these forces that were opposing him in the womb of the Macas forest.

"And what will you offer in exchange for this truth?" Kankiam asked studying the contents of Antonio's backpack scattered on the floor in the sand.

Kankiam watched the Shuara and awaited an answer that pertained to either goods or knowledge. He was surprised to see him fishing through his belongings where he retrieved a slender metal box that seemingly had no purpose or meaning. Necta held the box in the dim light and was careful to avoid sending out thought images about what he was holding. He could tell by the curious look on the face of Kankiam that his intuitive power was solely confined to mental images and not inanimate objects.

"Who is your most cherished family member?" Necta asked the startled panu statesman.

Kankiam was curious and his first thought response was, "Why do you ask," but he caught himself in time to know that if he had asked the question he would have broken shaman law by answering a question with a question. Not only would this answer have been an insult to his Shuara guest but also it would have been a horrible indiscretion that would have revealed his weakness as a seer.

"Chamik, my grandson," he verbally replied after hesitation.

"Tomorrow at first light if you will bring Chamik to me I will give you the gift of gifts. Instant images of you're loved one that you can take with you where ever you go. His beautiful image is always in your heart and mind, but with first light his likeness will always be with you in physical form on magic paper that only I can provide for you. This I promise to you."

Kankiam studied Antonio then answered, "Since you have given me an illusionary gift wrapped in a promise of tomorrow, then I will give you a half answer to your question wrapped in the promise of yesterday. You have given me an intriguing riddle and now I will give you one. The magic that defies and seeks you out is a untsuri Shuara shaman. Our mutual friend Achacho knows the answer to your questions. He would not tell you because you would not believe him. In your torn and troubled mind you are always seeking an answer from a higher authority. This is why you carouse with the whites. This is also why you are here. And because you are here I would be honored to have you spend the evening in my house where you will be safe. You can be content to know that at first light, when you present my gift of magic, you will have the power to defeat the demon that resides in the Macas womb. For me, I will be content to know that a miracle will bring me the gift of gifts. Good night and sleep well, my friend."

Necta knew he had crossed that invisible threshold that an uwisin could only dream about. He would be spending an evening in the personal lair of the most powerful shaman in the Amazon, which was an incredible honor and so prestigious in shaman circles that the wawek and Kakaram would easily sell their soul for. But, with this honor would come a sleepless night caused by concern that if he did drift into the sleep state the night stalkers that resided with and took instructions from the panu would compromise his most sacred secrets.

To thwart this problem, Necta reached into his medicine bag and removed a green quartz crystal and tied it to a small leather strap around his neck. He then reached into his jacket pocket and removed a silver cross that had been given to him by a friend, a Catholic priest. Kankiam had read him correctly.

He felt comfort in marshaling the good forces from all of his sources and this evening was no exception. He placed the cross near his heart for protection and reached for his green tobacco, which he rubbed into his eyes, knowing that the pain and irritation caused by the tobacco would keep him from falling into a deep sleep. His secrets would be safe this night.

The next morning when the sun was warm and the mist had burnt off the trail, Necta left the Village of the Shaman. Kankiam had kept his promise and armed him with knowledge and understanding of what forces were against him in the womb of the Macas Forest.

"Singusha is the one that turned his night stalkers against you," Kankiam had told him.

As Necta walked back to Wakani he had a new understanding that yesterday's magic was more powerful than today's. He knew that to ensure my safety on expeditions up the north bank he must find Achacho and find out more about his distant uncle, the legendary Jivaro shaman, Zamora Singusha.

He pursued his mentor and Achacho was receptive but, at first, hesitant to divulge any information. It wasn't until Necta explained the unusual behavior of the predators and his fear that a powerful wawek was undermining his efforts that Achacho finally open up.

Achacho told him the evil he was up against resided in a powerful magic from long ago that, through time and purpose, had grown in strength. This magic was directed and controlled by a deceased, yet still powerful shaman named Singusha, who succumbed to the dark side after his friend El Sapo was betrayed by the Colombians at Tambo Santa Rosa. The information ringed of truth for Necta. He recalled a confirming passage from the Mejia diary and how Mejia had befriended Singusha.

I welcomed Necta's pause to throw another log on the fire. My mind was racing; I was smack in the middle of the blurred line of reality and illusion. Necta's tale of the Village of the Shaman was very real for him and I accepted that. Then I remembered the strange feeling I had with Achacho and Necta when we stood at Singusha's old home site. At the time I had rationalized that perhaps I was up against a powerful physical opponent as Ireland had warned of. For the opposing force to be the Singusha from Mejia's last will and testament was, to put it mildly, a mind blower. I was trying to sort my thoughts when Necta delivered another super-natural mind bending blow.

Necta said he asked Achacho if it was possible to use his uwisin power to remove the curse on the land. He was told that Singusha still lived in the womb of the Macas Forest and that his spirit lingered near the remains of his house. The curse he placed on the land was not aimed directly at Necta, for he was a distant relative of Zamora Singusha. What he had experienced in the Macas Forest was only a warning from his uncle. If Necta were part of an expedition ruled by greed and betrayal then he would be held accountable by the powerful magic. Achacho had told Necta that the choice was his.

According to Achacho, Singusha was also an unta, and lived his earthly presence in both physical and spirit form. But, Singusha would never be called a "big man" because he switched sides from a pener uwisin to a wawek and directed anger and vengeance at all outsiders. He had warned Mejia to be careful with the green stones. I wondered if Singusha had utilized his shaman power to see into Mejia's future?

We had discovered in Mejia's last will and testament, his boss, a man named Barrana had betrayed Mejia when he sold Mejia's emerald sample and pocketed the funds. And the buyer of the sample, a British man named Perring had confronted Mejia and at gunpoint and demanded a map to the emerald source. Were these the actions that propelled Singusha into a mission of revenge? To protect the land from invaders whose heartfire was darkened by their impure thoughts? Achacho said that Singusha, with a heavy heart, retreated into the forest for a long time. When he returned he was drunk with a newfound dark power. Overnight the beauty of those times changed. Horrible deaths were attributed to the niawas that for the first time hunted in packs. The cascara dried up and the weather was unbearable. The Tungus they had fished and swam in was swift and impossible to cross. Singusha the unta-uwisin had gone rogue.

Singusha was just as effective in spirit form. According to Achacho, his power was birthed in hatred and revenge. His manipulative energy controlled weather and an army of night stalkers that preyed on unsuspecting prospectors and adventurers searching for the wealth of his domain. Everyone in and around the settlement of Sucua had heard the stories of the men that had gone up the river and never returned.

Did Necta's revelation surprise me when he explained how Singusha had taken it upon himself to protect and guard this area from evil? Or, that the bridge over the Tungus could not be built until the discarnate shaman gave his permission? A little, but then this wasn't a reality I was accustomed to. On some level of consciousness, I understood the endless delay in building a cable bridge after material had been delivered to Necta years earlier. I appreciated Singusha's—and Necta's—desire to protect this magnificent place.

Achacho had offered another clue: what Necta sought for his friends had twins. But he told Necta that now was not the time to discuss it. Then Achacho told Necta if Singusha approved he would divulge the location of the original Tambo Santa Rosa.

Achacho had decided before any natema induced meeting with Singusha would take place; Necta must send a telegram to Cuenca and ask his white friends to come to Wakani and deliver the steel cable necessary for building a bridge across the Tungus. The bridge was his excuse, a ploy, for the meeting to discover the heartfire, the truth to those searching for the green stones. If this meeting provided honorable answers then they would pursue the meeting with Singusha. If not, then Necta was told to terminate the relationship with us and return to his life as an uwisin. He would have no need to confront Singusha because Mejia's emeralds would remain safely concealed in the womb of the Macas Forest.

Necta said he struggled with Achacho's decision. He was apologetic when he said that he found it difficult to believe that his friends in Cuenca were manipulating and deceiving him for the sake of accumulating wealth. Dr. Vintimilla had been kind and showed respect by inviting him into his home in a city that frowned on such a gesture. I had given him friendship and trust. But, he also knew that we were human and susceptible to deception and greed. Before Alfredo, he had experienced greed in another form: men who used the guise of seeking sacred plants for medicinal purposes who had really sought locations of gold placer found in the rivers. Because of Achacho's concern, and under the scrutiny of Singusha's spirit, he had to make every effort to find the truth.

Thus, another piece of the puzzle fell into place. Singusha had offered me safe passage to Wakani. He honored his commitment by commanding the jaguar to break off the hunt of the tapir and for the brown snake lingering on the trail to leave. I now understood.

Necta spoke in a soft tone giving me the answer to the serpent and the bird. He said there were many spirits that roam their land, good spirits and bad. Achacho had told him that Singusha's command for my safe passage was challenged by an evil pasuk under control of a distant wawek to lure Esus to his death by the serpent. But, Singusha knew this and became the mighty bird that swept in and took the deceived reptile out of our path. He smiled when he said that even the Shuara must be extremely careful when something deadly whispers for them to come join in conversation. He said that Esus was young in spirit but in time would learn that evil was as strong and powerful as the good and this could be fatal, unless one has the special magic to see beyond the veil of deception.

Necta knew my mind was on overload and excused him self. I sat in silence trying to collect my rampant thoughts and emotions. When I finally came to my senses one thing was certain. Out here, we were indeed papier-mâché lions in a world of spirit guardians controlling and monitoring our every nuance and intention. I understood what the old prospector from El Paso meant when he warned me not to believe anything heard and only half of what was seen while in the Oriente. It was a true case of illusion versus reality and in this uncivilized land it was difficult to discern the difference between the two. I made a promise to myself that from that day forward I would not partake in any dialogue with a reptile or question spirit information or the existence of forest guardians.

The sequence of time moved rapidly around our primitive world. Wakani orbited the sun in sync with puffy white clouds and bright sunny days followed by decaying shadows that ushered in sequential nightfall. In rhythm with the cosmos a new year entered unceremoniously and soon dissipated into a blur along with distant recollections of my reasons for being here. I procrastinated about leaving the magical Wakani, but in the back of my mind time was slowly running out. I had made the decision to wait another week before telling Necta my search for Mejia's emeralds was over. It would be difficult to explain my change of heart, but I knew Antonio would fully understand the murmurs of my heartfire.

Sleep was a welcomed experience. My dreams or astral projections were active and I found myself floating above the triple canopies, observing the lay of the land in the Kionade. It was raw beauty and laced with many small rivers. I couldn't quite make out the mystical City of Magicians that I had heard resided in the Kionade and found myself questioning its reality as I sailed above the canopied forest. Had my extended circulation in the material world blocked my vision to the abode of the spirits? My journey of mind was my power dance. It allowed my soul to enter its own holographic image of experience that interacted with the harmony of my spirit—the necessary medicine for a wandering soul to find peace.

Every night at Wakani my mind and heart pulsed in rhythm with the soothing sound of the Tungus that allowed a deep, restful sleep. Then one night my slumber was abruptly broken when I heard Necta beckoning me to come quick and see a miracle in the making. I stumbled from my sleeping bag into the first rays of morning and was momentarily struck by the absence of sound from the Tungus. Overnight the mighty river had mysteriously slowed to a crawl exposing a riverbed never before seen.

We waded across the mighty old river that had successfully blocked our path all those years. My emotions teetered between euphoria and caution. Was I standing on hallow ground or partaking in a lucid dream? Was this a signal from Singusha that he approved my entrance to his guarded world? Was my decision to protect the Shuara and their sacred land from outsiders the stimuli that nudged Singusha to open the portal into the Kionade?

CHAPTER 16

KIONADE

The entrance into the Kionade

The Kionade was a mysterious place, a microcosm within a cocoon that combined beauty and danger and the ultimate challenge for any explorer. It was a journey into a land where time stood still. Necta's description of the Kionade jungle was hard to envision by anyone who hadn't been there. It was dense beyond belief with double and triple canopies filtering sunlight through a maze of flora. Once the trees punched through their thick terrace they grew tall and rubbed the sky with bushy leaves of emerald green. Kionade was a place of legends, a fascinating ensemble of raw truth and reverence. Hidden deep within this magical rainforest was a place the unta Shuara called the City of Magicians. Different from the Village of the Shaman, this mystical city was the birthplace of their tribal shamanism and they revered it as such. I wondered how much was mystical and how much was mythical?

When Necta described the land lying on the south bank of the Rio Tungus, the Shuara name for the Tutanangosa, his eyes grew wide and his voice high pitched. He had traveled there hunting for game and exotic plants. In daylight it was a true paradise, but at nightfall eerie foreboding replaced calm. That was when the predators left their dark caves in search of prey. Any man trapped within its perimeter at night was forced to bed down in the crook of a tall tree to keep him safe from the nocturnal crawlers like the Fer-de-lance and Shushupe.

Of equal concern were the four-legged predators, the jaguar and the mysteriously cunning and elusive black pantera that relied on a keen sense of sight and scent. After the wayward explorer strapped himself in a tall tree for the night his problems were just beginning. There were also those nasty little tree vipers whose bite was 70 percent fatal. They were abundant in the thick canopies, bathing in the sunlight by day and hunting nesting birds after sundown.

Similar to the Tungus, Kionade carried another one of those hidden meanings. A quick search in my Spanish and Quichua dictionaries produced no results for either name. Necta explained that Kionade translated to "rocks" and it was undetermined if this place of mystery was named after the terrain in the area or a physical condition in the brain for contemplating a trip there. What was certain was that it was the true womb of the Macas Forest of which Mejia had written. And it was also the green beast that devoured those that were lost and disoriented inside its invisible boundaries.

The Kionade was a dichotomy similar to a black widow's intricate web glistening in sunlight. It was intricate beauty, but for its next victim it offered a tangle of misery. Far beyond the lure of the emeralds a distant voice kept calling me to come see, come experience the hidden intricacies of this mystifying world. I was being drawn in like a pawn in a master's chess game and no matter how hard I tried it was impossible to ignore. I was surprised by my determination and refusal to let go. My mind was made up to venture into the Kionade, partly from my stubbornness and partly for the lure of the unknown.

Originally I had planned on a four-man expedition consisting of myself, Necta, Gonzalez and Macas. But since then Ambusha had surfaced and proved a trustworthy friend. He was a valuable addition to our team because he knew the lay of the land and like Necta, was a brave warrior steeped in Shuara tradition who roamed areas others refused to enter and did so without creature comforts and reliable weapons.

Ambusha saw and heard things that most of us could only conjure in our wildest dreams. The small bow-legged man was of average height and weight, for a Shuara, but unlike the others he carried an ominous presence. Anyone who wanted to harm him would never live to see tomorrow. He embodied the warrior image of the Untsuri Shuara and although he lived on the fringe of the Kionade it was obvious this curious man would have suffocated from within without the Kionade.

He lived near the Rio Mirruni, alone in a small thatched roof house far from the prying eyes of the colonials, the Catholic priests and undesirable members of his own tribe. He lived for the freedom to explore the unknown and the Kionade was his paradise. Yet he knew that to go there without permission invited death. Unlike Necta, who went into the spirit world to seek inner guidance, Ambusha did so for survival.

Ambusha was an arutam soul that walked the tight rope between civilization and his primitive heritage. His facial expression was almost always without emotion unless he was examining one of my shotguns or AR-7's, and then a thin smile would cross his lips. Unlike Necta who appeared younger than his years with eyes that reflected deep wisdom, Ambusha was just the opposite. His windows to the soul were stone cold black orbs that bore through you. His wisdom was carefully camouflaged and bound within a face that was ancient, a roadmap of a life of danger.

Gonzalez, in his twenties, was a typical young man with enthusiasm for life. Macas was probably forty and the mule of our group who relished testing his will and endurance by humping heavy loads under extreme conditions.

They were my team. The uwisin and his friends, one a mule of a man that trusted no one except Necta, the curious youth and a recluse that lived in a small thatched hut in dense jungle. Unlike Bill Wright who took a desperate expedition up the north bank with a small army and initiated balls to the wall approach, my team would fail or succeed through intelligence and patience. We were on our own, or so I thought.

The early morning sun peeked through the towering jungle as we left our safe haven at Wakani and waded across the Tungus. Within the hour we had cut a path upriver into the area that Achacho said was the resting place of the Tambo Santa Rosa. Necta halted our advance to scan the area for any sign of civilized remnants unclaimed by nature. There was no visible evidence of the old Colombian camp where Mejia and hundreds of other forgotten souls eked out a daily existence. We searched the underbrush and fallen timber for a hidden clue that would lay claim to the site, but the jungle had reclaimed its turf. The pace between death and disintegration in the rainforest was rapid. The trees lived out their lives and fell to their death to become food and the home to insects, spiders and snakes. Then the perpetual fungus took over, rotting the dead timber into submission until it became a part of the part consuming it.

Our search for remnants of the tambo ended when Necta and Ambusha climbed a small knoll to get their bearings. We were less than two hundred yards from Necta's ranch. I checked my watch and duly noted that from Wakani to what was assumed to be the tambo took less than thirty minutes to travel. Here we stood, just a short hike from Wakani.

We clocked the sun and pushed up the Cunquinza, also called the Flutes River. Soon, the large trees and mountains blocked sunlight and forced us to turn back. We took our time to chart the rivers, waterfalls and anything unusual that could be used as a marker. My compass, notebook and pens were the most valuable items I had, even above the weapons.

We were on a small knoll when Ambusha spotted a stream, slightly obscured by the forest to the west, flowing into the Tungus.

"Santa Rosa?" he asked.

We all looked at each other dumbfounded. Ambusha's question hit an exposed nerve. Before we headed up the river we assumed to be the Cunquinza, we had better make damn sure we were on the right river. Rather than group our forces we split into two groups to explore and determine the number of rivers flowing from the south into the Tungus. For the first time that I could remember, time was an ally and not a hindrance as long as we could make it back to Wakani before dark.

The sun was still high when Ambusha and I arrived at the stream with no name that flowed gently into the Tungus. My first thought was that it was too small, nothing more than a trickle. We barely got our feet wet. Undeterred, Ambusha was a man on a mission and his quick stride upriver forced me to scramble to catch up. The sparse jungle quickened our pace and soon we were in a flat swampy area of stagnant, slow moving water.

"Senor Lee, one cigarette?" he asked in halting English while staring at something a few feet away. I handed him a smoke. "Grand Bestia," he said pointing at a number of deep tracks in the mud.

"Ajiie," he said softly and pointed to a cluster of tracks scattered along the bank of the river. He then scooped up the sentiment-laden water in the palm of his hand for a taste. Scrunching up his face, he spit the water out and rubbed his lips on his sleeve. The water was tainted with salt.

I asked Ambusha if he were Mejia and responsible for bagging game for the camp workers, where would he begin his early morning hunt?

"Aqui," he said, without as much as a pause for thought.

I agreed. We were standing along side a natural salt lick that animals would flock to in early morning and evening for nature's mineral fix. If it existed in Mejia's time it would have been a hunter's paradise. Was this the place that Mejia killed the bear? Not likely! The stream's headwaters became a trickle and there was no river upstream to cross and no large rocks or steep cliffs along its bank. Was it possible Mejia tracked the bear from the salt lick and then made his kill on the river we crossed earlier in the day? Or, maybe there was another salt lick somewhere up water along the Cunquinza? The meaningful aspect of our speculation was that we were trying to get into Mejia's head and think along with him, after he awoke, had breakfast and loaded his shotgun to go hunting those many years ago.

While we examined the newly named Saladeros River, Necta's team was on a tedious search that turned up empty. There was only one river, the Cunquinza, according to Achacho, that entered the Tungus from the south between the Tunanza and the Najanbayhe. They backtracked to the mouth of the Cunquinza when the sun lowered its decent behind the thick growth and forced them to cross over to Wakani in shadowed light. Ambusha and I crossed back in early darkness and used the campfire from Wakani as our beacon.

The next morning at daybreak we crossed the Tungus single file and picked up our previous trail, marching with full stamina at a good pace toward our next challenge. The Cunguinza was calm and lying in wait when we approached. Before wheeling due south we huddled together to make sure everyone understood our plan. We would not split our forces. We would explore the east bank of the river searching for the rock open in the middle that Richard Ireland described as "a sultan's hat," and that according to Mejia was media quadra (50 meters) from its bank near a large landslide 400 meters in decent. The landslide might have disappeared over time but the jungle-crested ridge that contained the emerald outcropping would still be there.

The reaction among those unfamiliar with the Mejia passage was mixed. Macas seemed bored and more content that we were traveling light, leaving behind heavy tents and hauling only one backpack that contained food and water. Ambusha was detached and more interested in his AR-7 than all the emeralds the jungle might give up. But, everyone agreed we were on the correct river if Mejia wasn't delusional at the time he wrote his will. We would stay our course on the east bank of the Cunguinza and carefully pick our way through tangled undergrowth and thick foliage until we reached the first river upstream.

The five of us marched single file into a jungle teaming with dense overhang and towering trees. It became more difficult with each step and within the hour we stumbled upon a dry wash that might have been a river back in Mejia's time. Or perhaps not! It gave us pause for thought, which was rather academic, because we still had to cross and continue following the meandering river. We named the arroyo the Pequeno Obscuro (little obscured). I drew it on my map in green ink. Green indicated rivers and red was reserved for our trails, camps, waterfalls, and any thing else that I determined were unusual.

Our next obstacle was an arroyo that wouldn't qualify in present time as a river. Likewise this one was drawn in green and named the Rio Tsents. The name was taken from a strange plant that flourished along the far bank and was of particular interest to Necta. This was as close as my translation would allow since I had no Jivaro dictionary. Our snail pace was burning up daylight at a rapid rate and we knew that within the hour we would have to turn back.

Our allotted time was fleeting when we hacked around a bend in the river and came upon a sheer cliff facing that defied scaling. We had two choices; enter the pool in the river that appeared to be shoulder high or climb around through the green wall of towering vegetation and hanging vines to work our way back down to the riverbank. What began, as a walk in the park when we crossed the Tungus was now a laborious struggle to create a trail! What to do? Necta suggested we return to Wakani and the following day return and establish a permanent camp beyond this obstacle. This would give us a full day of exploring rather than our regimented half-day attempts and I agreed. Ambusha seemed detached and moody.

"Tomorrow we will be sleeping in the legendary Kionade," I said to my group.

"No Senor, Lee, this no Kionade." Ambusha corrected me. "Kionade numero uno is over there, across Cunquinza. Kionade numero dos, is farther. The Rio Cunquinza separates Kionade's. When cross Cansar we enter doorway to Kionade dos. To cross Cansar we must have permission from spirit guardians." He waved his hand in the direction of the opposite bank of the river we had almost forded.

"No go there without permission from spirits. No enter first Kionade without permission."

"Damn. Another set of twins," I grumbled.

Ambusha was adamant about his warning. He was acquiescing to the spirit forces and if they denied his request to enter either Kionade then he would not go. My mind reeled with questions. Did this explain how he survived all those hunting trips into the Kionade? Did Singusha grant us permission to be here by turning back the water of the Tungus? Was it a fluke of nature? Then I remembered he had used the word spirit in the plural sense rather than singular. Did this mean there were other spirits residing in the area that had equal or substantially more power than Singusha? It was a sobering thought that refused to slip away.

"How do we obtain permission?" I asked searching Necta's face for an assuring gesture that we were okay.

Unfortunately, Necta was noncommittal. The two men entered into a lengthy conversation in their native tongue.

Finally Necta relayed the discussion. "Ambusha's truth is his and in order to appease the warrior we must do one of two things to satisfy his spirits; take natema and seek permission face to face or visit with them in our dreams and they will either say yes or no. If they say no then he will not go with us tomorrow or the next day because he will be afraid. He also said that if they approve then you will be the first outsider allowed into Kionade without reprisal and this alone is a badge of honor and would qualify you to become a member of our people."

It was said I was the first white man into Little Hell and now I was on the verge of being the first outsider allowed into Kionade. Indeed it would be a deeply appreciated honor, but I was still thinking rationally and I wasn't prepared to indulge in mind-bending natema.

"Tell him, we will sleep on it."

In truth, my decision was based on the fact that I was a control freak. I wasn't thrilled about giving up my power in the middle of a jungle and go bopping into the world of waweks and Kakarams. I would take my chances with a bad dream rather than running screaming into the jungle at night under the influence of an unclassified hallucinogenic drug to search for the magical jaguar in order to become a man. Esus' talking serpent was still lodged in my gray matter and I wasn't quite ready to explore more of the unseen magical realms of the Kionade spirits.

That evening after our bellies were full and the fire withered away, I asked Ambusha how he knew so much about the area. He said that when he was a young boy, his father took him to a sacred waterfall that was a full day's journey beyond the confluence of the Cansar and the Cunquinza. As he grew older he was attracted to this place and returned every chance the spirits and rivers allowed his vision quest.

"This is place I spoke of earlier. You must go there, Senor Lee, not for green stones, but find yourself. It is womb where life began and to see self in present, you must return to the past."

With that prophetic advice my personal Kakaram had gained my vote for panu status. I nodded a silent response that he seemed content with and then he rolled over in his sleeping bag. Sometime later, deep in slumber, he began his power dance with his spirit entities and, hopefully, they would sanction he and his new sidekick's exploration to obtain their Arutam in the land of the spirits.

The next morning I held my breath in anticipation of Ambusha's answer. It was rather anti-climactic because he didn't go into details other than to say that permission had been granted and I should be honored. Curiously he was more sullen than normal and this played with my mind, but I quickly dismissed it. In the early light we strapped on our backpacks heavy with enough food, water and camping supplies to sustain us for weeks. The weapons were oiled and cleaned and the machetes razor sharp when we forded the knee deep Tungus and disappeared into a Marona stand of hardwood.

By mid-morning we were feeling the heat of the sun and sweating profusely under the tortuous weight of our packs. I would have given a hundred dollars for a caballo, a horse, to carry my heavy load, but the idea was insane; the horse would bring in predators and the poor beast would become nothing more than cat bait. The only thing to do was suck it up, keep pounding the jungle floor, knowing that every step was bringing us closer to the Rio Cansar, the gatekeeper. Was this what Ireland spoke of? "What you are looking for has to do with water...water by the gate or water by the door, water by the arch..."

We crossed the Obscuro and the Tsents and worked our way up and down the stone barrier encountered the day before. After hours of constant climbing and descending numerous canopy-covered ridges our progress came to a grinding halt. The heavy growth we'd hacked through had worn us down. Rest breaks were more frequent and my back and shoulder pain more acute. Noticeable angry red welts from the strap harness of the packs crisscrossed my shoulders. Beef jerky and hard rock candy gave us a brief but false sense of energy, and soon the weariness set back in and demanded the body stop the punishment.

On an energy scale of one to ten, Gonzalez and I were ranging between an unacceptable three to an average five. The rest of our team was in the high eights or nines and only drinking the fresh water sociably whereas we were living with a canteen to our lips. We were bearing down on the river named Cansar, which in Spanish meant to tire, tire out, and weary. It was aptly named. The sound of cascading water signaled our arrival at the doorstep of Kionade.

We cleared a small area that offered an excellent campsite. Dry wood was abundant and before darkness descended we gathered an ample supply for fire, cooking and, if required, security. The tents were placed in front of a large ring of rocks brought up from the river. When packed with wood and set on fire it would act as a firewall against any unwanted visitors. The hot coals would keep the snakes at bay. I was concerned about the cats.

"Niawas no here," Ambusha said.

The spirits told him the niawas would stay on the other side of the Cunquinza in Kionade numero uno, and would not bother us unless we entered their domain, and then we would be at their mercy because they were hungry.

"We no go day or night," he said convincingly. My resident panu was dead serious. But, something else was still bothering Ambusha. He was detached and surly.

Huddled in the fading light, Ambusha and Necta helped me coordinate my map. Ambusha drew from memory of his past experiences in this area. We were camped at the confluence of the Rio Cansar and the Cunquinza and according to Ambusha we were on neutral turf and must not move our camp any further. Once we crossed the Cansar we would be entering the second Kionade.

I wondered what Ambusha's spirit friends said about our expedition entering the second Kionade. I asked Necta to discuss this issue with Ambusha. That was a mistake. Their native tongue discussion seemed to go on forever and appeared to be developing into a heated argument. Finally, Necta halted the give and take and said simply. "Ambusha said that the spirits told him not to enter the Kionade across the Cunquinza, but we could enter the other Kionade across the Cansar, but not camp there after dark. It would be too dangerous. He did not say why. He said they told him that we must not enter with an odd number above one. Three, five, seven or nine are forbidden."

Ambusha was fanatical about spirit numerology. It made no logical sense but after what I had seen and heard in the Oriente there was no argument from me. I wondered why he didn't tell us before we left Wakani that we were operating under an odd number curse? There were five of us. When Necta scolded him for allowing this to happen, Ambusha defiantly said he was told by the spirits not to divulge this information. However, since we were all his friends he felt he should warn us. The spirits would make final determination who continues and who returns to Wakani. If he'd told us earlier then Necta would have made the decision for them and this would have interfered with their plan. I was concerned who would be the first to leave and what was their plan?

If a man wanted to exercise belief in his faith, this place had to be the ultimate chapel. The impenetrable personal privacy that we human beings believed was our spiritual first amendment was a myth here where spirits ruled and dictated one's every move. However, Necta still believed in his truth. His faith leaned on Singusha, his discarnate uncle, as the controlling spirit of the unseen forces at work. My truth hovered somewhere between both perspectives. I found myself leaning toward the side of Singusha, only because he had demonstrated his power by drying the Tungus and the fact I hadn't seen any fanged predator in over a month.

I found myself listening to and agreeing with Ambusha to explore the east bank during daylight and retreat to our neutral base camp before nightfall. We would be cutting into our exploration time but by doing this we would pacify Ambusha and the entities that had taken up residence in Kionade two. Down deep inside I hoped that Singusha could control any competitive wrangling within the spirit world should it arise.

I wondered if I would have been receptive to continue on had it not been for my early upbringing in my grandmother's Pentecostal religion? I had been conditioned as a youngster to believe in spirits and observed her on numerous occasions speaking in unknown tongues, the language she called the 'Holy Spirit'. These strange, but real experiences, along with my spirit dog at Infiernilios and the discarnate energy in Adriano's house had solidified my belief that an unseen world was encased within another dimension surrounding our everyday life. Without those reassurances I probably would have labeled the two Antonios crazy and made a hasty retreat back to the sanity of my material world. It was obvious that in the world of the Shuara, one must become like them or leave. I opted to stay.

CHAPTER 17

POWER DANCE

The abode of the forest guardians

The next morning, Gonzalez came down with a mysterious intestinal bug that left him dizzy and too weak to travel. Necta seemed relieved. He was the youngest member of our team and the ideal candidate to be left behind. Gonzalez was the first odd man out. Ambusha's spirit guardians seemed to be wise in their choice because we needed the mule to carry the heavy loads the healer to maintain our health, our guide to guide us and, me, the riddle solver.

Safely entrenched in an even number we waved farewell to Gonzalez and shoved off into the cool mist rising from the river below. For the first time I saw clearly what had lulled me into a peaceful and restful sleep. Three waterfalls similar to stepping-stones formed at the confluence of the Cansar and Cunquinza. Was this the "entrance by the cascabel," that Mejia described? Lost in the noise of cascading water we cautiously probed the first fall and managed to climb the slippery slope to the second. With caution and determination we pushed upward until we reached the last barrier. It was here at the watery doorway to the second Kionade that our trouble began.

Macas lost his footing in an attempt to navigate around the swirling current and was swept away into a pool of deep water. We were powerless to do anything except hold our breath and wait. It seemed like an eternity before his head popped up. He took a deep breath of air and went back under. Macas was strong and powerful, which was evident when he resurfaced with a heavy waterlogged backpack in one hand. Ambusha worked his way back down the slippery rocks and helped drag our mule to a shallow area. He seemed more embarrassed than concerned about the bloody gash in his leg.

Necta and I waited for our two compadres to join us. Suddenly I was struck with the odd feeling that something or someone was watching us. Perhaps it was the dampness from my clothes, but a chill raced up and down my spine and triggered my ready alert system to stay vigil. About the time calm was restored with my thought processes, Necta brought my attention to a group of monkeys silently staring at us from tall trees across the river. Below them a large pile of driftwood and tree trunks formed a natural bridge over the river into the forbidden zone.

Necta's eyes betrayed his thoughts. Was this a test to see if we were going to honor the spirit's request and not cross over into the first Kionade? Oh, I wanted to test them as much as they wanted to test me. It was tempting and frustrating to see pasture-like terrain without any thick jungle for at least fifty yards beyond the river's bank. It was tempting, because Kionade uno would give us an advantage in time and clear observation of the area we intended to search. Instead looming before us was a maze of tangled undergrowth and thick trees. Surely the forest guardians wouldn't take offense if we entered briefly? Then again, were those silent watchers in the trees acting as sentinels? Unless I was reaching, it appeared the nature-created bridge was the carrot being dangled to test our truth?

Ambusha carefully eyed nature's bridge and our reactions. A thin smile crossed his lips and then without hesitance he ignored the bridge and headed up the Oriente (east) side of the river. We followed close behind into a green blur of thick jungle dominated by broad leaf plants and hanging brown vines when I detected quick movement in the upper branches.

"Culebras," both Antonios said. They warned me to be careful of a deadly brown tree viper that took on the appearance of an unassuming vine. If that wasn't enough to worry about there was the Loro-machacuy, a green eight-foot tree climber, and a plumb colored tree viper. Unfortunately, everything we chopped through or that dangled above us were these colors. The tree vipers were restless. They moved about in the upper canopies jockeying for position in the warmth of the sun and seemed uninterested in the movement below.

On occasion we would chop into a small clearing that allowed a respite from anxiety and the opportunity to wipe sweat from our brows and eyes and take welcomed breathers. During these brief breaks we sometimes found ourselves close to our river and other times far from it. Never once did we seem to have the luxury of traveling in a straight line; instead jungle obstacles dictated our path and we always chose the one of least resistance.

I understood why Necta lobbied long and hard from the very beginning that we plan on a lengthy stay if we were to be successful. We could have been within twenty feet of the sultan's hat and never saw it. Finding a ninety-two year old landslide reclaimed by nature was equally daunting. Mejia wrote that the rock open in the middle was media quadra from the river or about 50 meters. It was impossible to judge distance in the dense growth and we were like four blind mice hoping the cats were still asleep.

The afternoon ticked on. We slumped from exhaustion against some tall leafy trees that Antonio pointed out as cascaras. These were the source of quinine that had attracted the Colombians into the forest in the previous century. Necta said an old Shuara pener uwisin discovered its use and treated tribal members for malaria long before the colonials came. He pointed out a strange looking bush he called Malva whose leaves, when boiled, was an effective heart stimulant. There was a tree named Leche de Oje whose prepared sap provided violent purging of internal parasites such as worms and amoebas. Necta was in seventh heaven knowing that these remarkable cures were less than a day's walk from Wakani providing the Tungus stayed shallow. Then he approached a tangle of brown vines and with his machete cut them in twelve-inch lengths.

"Ayahuasca" he said and placed them in his backpack. Finally, I had seen the "vine of the soul" in natural state.

After my brief pharmaceutical lesson we plowed back into the dense growth that knew no boundaries. Thick creeper vines brought out the machetes and chopping consumed valuable time. Bird life was more abundant and raucous calls greeted us as we penetrated deeper into their sanctuary. Were they applauding us for being true to our word or mocking us for being such fools on a difficult path?

During lulls in their chatter we heard running water. It seemed to be coming from the south and we assumed it was the Cunquinza. But, this posed a problem because Ambusha's map indicated the Cunquinza made a dramatic turn beyond the sacred waterfalls and headed due west. We decided to investigate. As we closed the distance the more disturbing the sounds became. One minute the flowing water was detected south of us, the next minute it sounded like it was west. Like the mysterious bellbird that decoyed men to their death in the wilderness by luring explorers in circles with its haunting sounds, the surround sound of running water left us confused and disoriented. The omni-directional sound of the water played havoc with our sense of direction.

We climbed a ridge that offered a partial view of a large bend in the river. This was where the mystery river changed its direction and flowed parallel to the Cunquinza. The ridge on which we took our rest break straddled the two rivers. Ambusha pointed in the distance and said that further up the river several miles there was another large waterfall and a river his father called the "Cascabel". My mind raced. Were we only a hike away from pay dirt? Was that the place where Mejia found the emerald deposit?

Although my mapping was more accurate and complete, our curiosity cost us dearly. We had pushed in too deep, lost track of the sun's position and the needed hours of precious return time. It was impossible to return to our neutral base camp before nightfall. To the dismay of Ambusha, we had to violate his 'friends' edict not to spend the night here and he was concerned that trouble was on the way. Was there something profoundly sinister rallying its forces, waiting for the magic hour when we set camp at night in Kionade dos without tents, warm food or shelter? We were all trapped in Ambusha's nightmare.

The last thing we wanted was to spend the night out in the open with four directions to defend, so in the remaining light we made a beeline for the Cunquinza. A camp in front of the river would offer an abundant supply of driftwood and fallen timber for a firewall and one less position to worry about. After an hour of cutting a trail we reached its boulder-strewn bank. Necta and Ambusha combed the bank searching for a natural cave or hanging rock that would offer protection for our rear flank. Macas and I chopped and stacked dry wood on a small playa that offered more protection than the mountain ridge we had descended.

Suddenly Necta's high-pitched voice ordered everyone downstream to come at once. He had found a small cave chiseled by nature into a ledge a short distance from the river. Our firewall around the cave took on new meaning but opened debate between the Antonios. Necta proposed that we light it and keep a roaring fire going through the night. It was doubtful any of us would sleep anyway with the exception of Macas who was running a high fever from an infected gash in his leg and required sleep. With everyone else on high alert the fire could be attended to without fear of it dying out. This would keep the night rovers at bay in particular the deadly fer-de-lance. There were obviously a dozen more nocturnal types, but for obvious reasons the one previously mentioned stuck in my mind.

Ambusha opted for a fireless evening citing his concern that the glow from the fire would attract anything and everything curious to our front door step. "Let's sleep in the trees." Ambusha said.

The argument between the two was healthy and informative. Necta agreed that climbing trees would protect us from the niawas, but not the tree-hopping vipers stirring in the towering forest above us. Ambusha said he would rather take his chances with the culebras than the niawas.

Unfortunately, I was chosen to break the deadlock and after considering both opinions carefully my decision came down to Macas and his deteriorating physical condition. His eyes were watery and he was running a high fever. It would be impossible to lift him into the top branches of a tree and even if we were successful, how would he defend himself in a fevered state of mind? If we stayed near the river in the small cave the dampness would compound his chills after the fever broke, but at least we would be at his side to protect him.

I voted we stay where we were and light the shield of fire only when Macas began to experience difficulties. We would be exposed and forced to rely on our senses in the darkness, but with a slight breeze idling from south to north we still had the edge. If the wind changed directions and carried our scent across the river we would be forced to light up our protective wall of fire.

Ambusha relented. He was a warrior, but even warriors had a weak link in their armor. His was the niawas. Necta's was the snakes and mine were both of those and every other creepy and crawly thing in between. When you really got down to it, did it matter whether you met your maker in a fast and painful death with poison flowing through your veins or a slow and horrific gnawing death with a jaguar holding you down with one paw while he enjoyed a casual lunch? It was going to be a long night.

From our makeshift fort the view across the river was spectacular. Trees were so high they embraced the low hanging clouds, their trunks so sturdy that if a man could carve out their innards he would have a room with a strong roof and sturdy walls. The triple layered jungle terraces were home to monkeys, assorted bird life, tree vipers and flaming orchids reaching for sunlight. Broad leaf plants, large and small, mixed with the tangled vine undergrowth. A picturesque sandy playa separated the heavy growth along the riverbank and offered a clear view or shot if niawas decided to swim the black inlet pool of slow still water and charge us head on. The logjam upriver proved beneficial because it had created deep pools.

Shadows replaced sunlight when Necta and Ambusha hurried into the jungle in search of the Copaiba tree whose sap was excellent for curing open wounds and diseased areas. The Curarina bush was also on their pharmaceutical shopping list, it was used by the uwisin to treat infections. I had left my antiseptic and medical supplies in camp. I thought it would be a daylong outing, not an overnight in the forbidden zone.

I took my buck knife and cut an opening in Macas' pants leg that ran from the cuff, to just above the knee. A deep gash some five inches in length ran from the knee down the calf of his left leg and the wound looked angry and swollen. The jungle swarmed with bacteria and I was concerned that it would be only a matter of time before his wound became infected with gangrene. I emptied my pack of hard items and propped his head between it and the rock surface of the cave wall. I lit a cigarette for him and he took a deep drag, leaned back, forced a toothy smile and apologized. I shook my head and smiled back at him. There was no need for an apology.

Necta returned with a hand full of herbs he called Ayamullaca and mixed them with water to make an anti-septic soap used to cleanse wounds and infected areas. I poured canteen water into Necta's cupped hand, we mixed the two together, and bathed Macas' injured leg. He winced from pain but never said a word or pulled away and continued to attempt a smile through dry swollen lips and a fever induced haze. The Shuara, like the Apache, had a threshold for pain that was unbelievable. I held the canteen for him and he took a long deep drink of the liquid of life then lay back down and closed his eyes. In the calm Antonio asked, "Lee, do you fear Ambusha's spirits?" I shrugged my shoulders in uncertainty. In truth, I was apprehensive as hell.

I returned his question. "Are you frightened?"

He shook his head. I asked if he was frightened for Gonzalez, all alone at our base camp, a full day's walk from us. He responded quietly that he wasn't afraid, because he believed the forest guardians would honor their commitment to keep that area neutral. He was concerned, but not fearful. He measured his words carefully. He was concerned about Ambusha's safety since he was their messenger and responsible for keeping us out.

I asked how the spirits would react to his failure?

"Singusha must keep the balance," Necta said. "This will be difficult because Ambusha's fear will attract others with their night stalkers. Fear disrupts the harmony of everything living in the forest. Like the culebras, when they detect the fear, they strike. I learned my lesson that evening the night stalkers struck my camp on the Tungus. Had I known what I was up against, I would have prepared my defenses differently."

I asked if the night stalkers were real flesh and blood or were they an extenuation of imagination brought on by the fear he spoke of. His answer was chilling.

"Lee, they are as real as you and I."

He was implying that the jungle or la selva, as he preferred, was a microcosm of unseen energy that thrived on the rhythm and interaction of life that existed within its boundaries. This included plants, rivers, insects, snakes and all wildlife. This force recognized any disruption in its natural life cycles brought on by intruders. Man was definitely a disturbing threat to the ebb and flow. Greed, anger and fear brought disruption to the microcosm. Our problem was our thoughts and how Ambusha would deal with his guilt, anger and fear. Necta had to come up with a spiritual magic bullet to release his anxieties. If he couldn't then the rain forest harmony would be fractured by Ambusha's emotional upheavals. He was torn between his loyalty to them and the reality of our predicament.

Ambusha was withdrawn, anxious and controlled the conversation for what seemed like a long time until Necta began another one of those singsong conversations that captured the warrior's attention. When Necta finished they embraced each other and Ambusha seemed more at ease with himself. My god, I thought to myself. What did Necta say to him? I couldn't wait for Necta's translation.

Ambusha's fear was warranted because of a previous close call that prompted a promise to his deceased brothers that he would never return to Kionade without their permission. He was frightened because he had once again broken a promise and found himself at their mercy.

Necta consoled Ambusha. He explained how the threat of death brought men to the threshold of faith in their Great Spirit and asked Ambusha if he believed in a God. This struck me as odd for I had never heard Necta refer to God before. Ambusha answered that he believed in a spirit force. Then Necta asked him who or what controlled the spirit force? He couldn't answer. Then Necta explained what he had learned from the ancient ones and pointed to the spider web in the corner of our cave and explained the connecting patterns that formed its usefulness. It was the old Shuara belief that creation existed in everything and was connected to a rhythm of all creation whether it was a human, or plant, or rock. The spirit guardians were not gods or the final solution because they too had to answer to Creation; they were only a part of the system.

Necta begged Ambusha to bury his anger, fear, and guilt that he had failed the forest guardians. If Ambusha's spirit friends had honorable intentions they would realize he did not intentionally violate their edict. Our placement in the Kionade was an accident and not because of anyone's defiance. He concluded with, "We must all be as one in the harmony of nature and its creative force to survive this evening." Ambusha understood and nodded in compliance.

Necta warned us that in spite of our attitude change they would still come around midnight in the first light of the moon to test our will and courage. "There will be a power dance between those that seek vengeance and those that seek harmony and we are not to interfere and take life unless we are absolutely forced to. Just be prepared."

In my worst nightmare I couldn't conceive of what we were up against. My skin crawled with anticipation.

When the last vestiges of sunlight disappeared into the shadows of darkness a strange calm came over the dark world blanketing our makeshift fort. In front, a 30-degree slope ran down to the riverbank some forty feet. In the rear and on top of the small cliff the tree line was dense, but there was no overhang of branches that would allow snakes to drop in on us. A six-foot pile of firewood was stacked to the side of the cave and blocked out one angle to our nest. A ring of rocks knee high formed a circle perimeter around our position. In the center and along the edge of rocks sat a pile of deadwood and dry grass waiting to be lit. We were prepared in the event the power dance ended in our disfavor. If the night stalkers came for us they would have to enter a wall of fire and firepower to breach our defenses.

Necta and I took our positions on the corners of our crevice with automatic shotguns and Ambusha sat in the middle with his AR-7. Macas was asleep, his mangled leg stretched out between us. We had forty rounds of 00 loads sitting within our reach in bandoleers including over a hundred rounds of hollow points for the AR-7 rifle. We also had two flashlights with extra batteries, two miner's picks and a bag of piri-piri that Necta admitted only worked with certain snakes. Experimentation with cattle bitten by the colubrids and their subfamilies proved they were exempt from the jungus plant. But, for me it was still a psychological boost to know we had anti-snake venom that didn't require refrigeration.

Our food supply was down to a few mangos and several boxes of saltine crackers that for tired and hungry men tasted like filet mignon. Our original plan was to be back in camp for our evening meal instead we were hunched down in a crevice with cradled weapons squinting to adjust our eyes in the blackness of night. My body was numb from the physical battering of the trail and my mind wasn't doing much better. I longed for sleep. It was impossible to focus on the black void across the river.

My thoughts wobbled back to the dam of fallen logs downstream. It was a double edge sword. Both Antonios knew the natural bridge was a test to see if we would honor our commitment and stay out of the old Kionade. I wondered if since we had honored our commitment was our repayment the excellent defensive position? What roll was Singusha playing? Did he guide Necta to the cave and the deep-water barrier? This entire expedition was a lesson in illusion and reality in this emerald green Land of Oz.

A slight breeze carried in a sickening sweet smell of decaying plants. In spite of the nauseous stench it was good to know we were upwind from the niawas. I checked my watch. The next couple of hours I struggled to keep from nodding-off. In a way I wished it was thundering and raining to break the monotonous nerve-racking silence of the darkness. We continued to hunker down and wait for God knew what. I looked to Macas and the two Antonios for reassurance. I saw darkened frozen forms.

Sometime around midnight the first glimmer of the full moon peeked through the tall trees and reflected an elongated light in the still black water beneath us. This was the signal Necta had warned us about. It was time to double-check our weapon safeties and locate our spare bandoleers of ammo. I surveyed the men and our position. We were like a night patrol hunkered down in a bunker waiting for an enemy to cross the river before we opened up with lethal firepower. The only problem with this analogy was that according to Necta, the enemy was within us.

Our attention was directed to the tree line across the river when a diffused light appeared. It was without solid definition. It moved from one tree to another, slowly working its way towards the playa, stopping then starting in a graceful and intelligent manner. It appeared to be examining bushes, large rocks and fallen logs.

"It's a good forest spirit examining the health of the forest." Necta whispered. "It means us no harm and is only doing its job."

The light appeared to be more like a concentrated ball of energy with intelligence behind its movement and action. It moved slowly along the riverbank then stopped and circled near a large boulder and vanished. I took a deep breath and wondered if we were all hallucinating.

Within minutes another diffused light ball appeared in the tree line followed by others that slowly levitated towards the river. I counted three and eventually four that moved in a graceful manner stopping in unison every so often as if to scan the terrain. Necta appeared to be uttering a silent prayer. But to whom was he praying? Was he speaking to them or maybe Singusha? Then he reached over and put his hand on the shoulder of Ambusha and whispered something to sturdy the troubled man.

"Energia baile" he whispered to me.

The energy-power dance began when four ghostly images made their way towards the black lagoon, a stone's throw from us. I couldn't believe what I was witnessing. My finger tightened on the trigger then I thought how stupid I was for thinking I could shoot a ghost.

Suddenly their movement stopped. But why? Was our side of the river off limits to them?

When the unworldly gathering paused just short of the opposite bank my scalp became taut and the hair on my neck and arms felt tingly. My body's strange reaction signaled my brain to take heed and in spite of Necta's warning fear rushed in. In the past fear had nudged my senses but vanished when I took control of my own destiny; however this time there was a different set of rules. I was in a mental struggle with a primitive energy force that had tapped danced with my subconscious and toyed with my mind.

I was not alone in the struggle. Macas was conscious and propped on one knee. He stared in disbelief as the energy clusters formed a rank facing us. The two Antonios had been through this before and seemed relaxed with the ghostly dance, but Macas and I were amazed. My eyes strained to make out a human form, but there was none. Instead, there was only diffused pure light that remained consistent. To me they were ethereal forms of geometry and intelligence that appeared as mediators for a sinister presence awaiting its entrance.

Necta schooled Ambusha, "Blend in with nature and creation and think in unison."

Sounds easy, it wasn't. The man hunkered next to me had white clenched fists and eyes wider than mine. He had been asleep when Necta gave us the spiritual pep talk and had reacted like any rational thinking human being would under extreme duress. How could we find a neutral zone and chill out?

Humor! This distraction proved reliable and sustained me in the past. If convincing, it could repel any negative energy feeding off fear. I nudged my worried friend on the shoulder to get his attention and with a chuckle and a calming voice said, "Hermosa hadas, si?"

Then, after what seemed like an eternity, he computed my strange comment and finally responded? "Si Senor Lee, muy hermosa hadas." He relaxed. I too relaxed and offered another quiet chuckle to punctuate my stupid, but effective comment that the images were only beautiful fairies. Hopefully we were all sending positive thoughts to this truth squad floating above the outer bank.

Within minutes our neutral zone was breached and put to another test. The stillness was shattered by shrill sounds similar to high-pitched flutes, off key, but in unison the noise surrounded our position with nerve throbbing shrieks. The full moon was high and for the first time we could see forms and movement in the treetops, but it was impossible to try to pinpoint the sources of the flute sounds. Necta shook his head. He had heard this before.

"Verde Loros," Antonio whispered. The tree-climbing parrot snakes that searched and destroyed nesting birds in the trees under the light of a full moon were rallying their species. The Shuara believed the shrill sounds signaled each other the positions of the defenseless parrots, macaws and numerous other species sleeping peacefully in the trees. I understood why the Rio Cunguinza was called the Flutes. The large trees, zigzagging along its banks, were gigantic nests for one of the most vicious tree-climbing vipers in Ecuador. We had camped in the heart of their territory against the best wishes of Ambusha's forest guardians who lingered nearby monitoring our fate.

I momentarily craved nicotine to calm my nerves, but knew that a match strike might bring something unwelcome into my life. The moonlight now bathed the jungle. A surreal scene blended into sounds of horror from the Verde Loros feeding frenzy and their victim's final haunting cries as defenseless birds succumbed to the vicious attacks of the vipers. The horrific sounds of those dying and the wounded begging for life assaulted my senses. Sweat rolled down my forehead and slid into my eyes with a stinging sensation. I closed my eyes and hoped to awaken from a sleep-induced nightmare; instead the power dance continued.

Across the river in the pale moonlight I saw black silhouettes—swarms of birds in the safety of flight. For some, it was escape, but to others that had been slightly penetrated by the lethal fangs dropped from the sky like heavy stones. In the jungle behind our cave the screeching left very little to our imagination as gravely wounded parrots fell into the river. Moonlight illuminated a horrific scene. A large parrot flapped its wings in despair, crying for help, immobilized while a slender snake held it in a death grip. Dozens of wounded birds fell along the playa flapping their wings in final death throes.

Beads of sweat poured into my eyes. My throat was dry; it was painful to swallow. I wanted to cry, help the birds, scream and run to safety but there was no safe zone in this horror show. I glanced at Macas and then Ambusha. In the moonlight their silhouettes seemed stiff and lifeless frozen in stoic anticipation. We were all lost in our own worlds of doubt and surrender. We were at the mercy of nature and she was undertaking a cleansing that I realized was natural and with purpose. Still I wondered if we were next on the hit list of nature? Or even worse, was there a disgruntled forest guardian or wawek in control of our destiny?

The snakes were close. I had reached my breaking point when the moonlight slowly vanished behind the tall rainforest. I stared at Necta hoping to get his attention.

"Light the fire Necta. Please light the fire!" I yelled over the din. Then I froze.

A slender snake glided by his shoulder near the stacked firewood and then disappeared into the darkness. I swallowed hard and breathed shallow afraid to move. I fully understood the meaning of "frozen in fear." Humor where were you?

Necta either read my mind or heard my plea and he too had seen enough. The fire crept slowly at first then fueled by dry branches and small sticks raced through the wood and within minutes the flames were knee high gaining strength and intensity. A slight breeze funneled the smoke upward in a swirling pattern among the treetops inducing panic among the tree vipers. When the moonlight reappeared I detected furious movement in the dense canopies from the vipers escaping their worst enemy, smoke and fire. Their eerie whistling sound was moving away and indicated a retreat was in progress. The relief was momentary.

Without warning the wind shifted and the dense smoke billowed into and clung to the walls of our makeshift fortress, stinging eyes, filling nostrils and creating coughing jags. We had no choice but to move out of our bunker into fresh air. Were the guardians angry with us for tipping the balance of their power dance? Were they literally smoking us out? I helped Macas to his feet and moved him beyond our ring of fire. We reached the riverbank and formed a semi-circle coughing up smoke, gasping for fresh air, our watery eyes scanning the ground for any sudden movement.

The moon slid slowly behind giant cedro tree branches. Its brilliance diminished too soon for Ambusha. He detected movement across the river near the tree line and this time it appeared to have size and mass.

"Niawas," Ambusha whispered.

Four unidentified dark forms moved from the tree line toward us. We quickly retreated back into the safety of our cavern and took up our previous positions.

As suddenly as it began, the power dance ended when the shadowy forms disappeared into the darkness of the jungle behind them. The light sources glowed brightly for a few minutes then dissipated into darkness. We all breathed a sigh of relief and lit up. I checked my watch and it showed a little after 2:00 a.m. For the past twenty hours we had been on an adrenaline rush with little food, rest or sleep and my body and mind were spent. Only four more hours of darkness remained before we could escape this horrible place. Then I nodded off momentarily.

Sudden movement and a cry of alarm jarred my dulled senses. Ambusha was jumping, slapping at his leg trying to dislodge something from his calf. I fumbled with my flashlight thinking that a palm viper had somehow managed to penetrate our firewall and had buried its fangs into Ambusha's right leg. I switched on the light and saw movement on the cave floor. The intruder was a large reddish-brown spider scurrying for the safety of his web in the corner. Macas took a rock and smashed the eight-legged spider into submission. Necta quickly examined the bite area, but there was nothing anyone of us could do. I gave my friend a coveted American cigarette; he mumbled thanks and then something in Shuara.

"He believes he will die from the poison of the spider bite as punishment for defying the spirits," Necta said. "But he is prepared to accept this without anger or remorse and only asks that we bury him in his house alongside the Rio Mirruni. Lee, there are so many spiders in this region and many are poisonous and others harmless and to have knowledge of all these things is impossible. I will try and encourage him to live, but first he must dispel his thoughts of punishment. We must wait one hour and see if he wants to live. If he agrees, he will only be sick for a few days. Then at first light I will find the copaiba tree and speed his recovery."

Ambusha allowed himself a small smile. I gave thumbs up to the brave man who was in his mind teetering at death's door. He acknowledged me with a smile and a cigarette dangling from his lips. Necta wiped his brow and whispered broken syllables in their dialect. It was a timeless incantation that calmed our wounded comrade. There comes a moment in life when nothing else matters except life. The hour passed and Ambusha was still alive. He and Macas fought pain and uncertainty in their own ways and neither succumbed.

Courage and strength were overused words out here and never seemed to aptly apply to any given situation. I was struck with how easy it was for these people to surrender their life force to tribal beliefs and primitive folklore. Death was only another doorway to a new path of adventure and fulfillment of their 'ancient specter' soul. Ambusha wasn't concerned with his death only that he failed his spirit friends. Absent of the comforting belief in a Supreme Being, his faith centered on a form of reincarnation. He believed he would return in spirit form to a little thatched hut along side the Rio Mirruni and join deceased loved ones and they would continue to hunt and fish behind the veil of this new, but old world. The indigenous people of the Amazon after a life of hardship welcomed the after death experience because they had nothing to lose.

Our team was slowly being whittled down. Macas was without the full use of his left leg and Ambusha his right. Unfortunately this meant a hobbled retreat to our base camp come daybreak. My thoughts went to Mejia and Perring and their respective loss of a leg and foot that had prevented them from reentering the Macas Forest. The spirit guardians had aimed their tsentsak well.

At first light we struck into the forest and after a brief pause Necta rejoined us with a gooey white sap from the Copaiba tree. He mixed it with water and applied it to the infected areas of our wounded comrades. We cut two strong limbs from a tree and fashioned crutches for them and then headed toward the Tungus. Necta took up the point and I the rear. We unlatched our safeties on the shotguns and waded into a wall of green without sound. The birds that greeted our arrival the day before were gone—dead or silent and the stillness added impetus to leave this area as quickly as possible. We had lived on the edge for the past twenty-four hours, survived a power dance, night stalkers and limped to safety.

The sun dropped along the ridgeline and formed gray shadows as we edged closer to the Rio Tungus. My thoughts went to the cascabel and the rock open in the middle and how close I might have been to the emeralds resting place only to discover they were a part of the energy of the land. To remove them required permission from the forest guardians. It didn't seem right that Mejia played out his destiny in life by leaving a treasure that attracted and compelled ignorant souls to enter this no-mans land ruled by discarnate energy long passed but not forgotten.

The slumbering Tungus seemed to welcome our safe return to Wakani. Later, in the stillness of darkness my thoughts went back in time to a tow-headed youngster that had began his expeditions in the dry washes of the Apache Reservation and culminated years later in the womb of the Macas Forest. It had been one hell of a ride and now I felt it was over. But before I left Necta pulled me aside. He said the spirits told him my return would be a long time in coming, but I would be back.

CHAPTER 18

ABSOLUTION

Antonio Necta and Lee Elders

Time had marched on since I last set foot on Kionade soil. Adriano and I had stayed in touch but the adrenaline pumping letters of yesteryear quickly faded into the reality of my career moves, marriage, family life and personal responsibilities. Occasionally my heart would flutter from his disclosures of new finds in the Oriente. But my interest waned due to my self-imposed exile from the dangers of nomadic adventure. I had overcome my addiction to adventure and its twin, danger, through a smooth transition into the exciting discovery of a rewarding life.

Then it began.

Dreams roiled in my mind, breeding and living through their own primitive design. At first they were only wispy and hazy images with shades of green foliage and white puffy clouds in deep blue sky. In the beginning their only purpose seemed to be sleepless nights and curious hours of reflection with my wife, Brit, and our son, Brett. They were both amazingly supportive and suggested that I might consider a return trip. I vetoed the idea because I was busy living in the real world and couldn't take the time to wander off to the jungle.

But the dreams progressed. They seemed to strive for integrity in their birth, informing some level of my subconscious that something new and unpredictable was about to happen. They took on a dimensional effect and, for the first time in my life, sound accompanied the visual imagery. Over the roar of the Tungus I heard my friend Necta begging me to come to Wakani.

"Lee, Lee," he cried out. "You must come and help me. You are the only friend I have left. The others are gone. Please come."

In one hand Necta held gold and in the other a large green stone.

Then his image and conversation faded into the void of sleep and I would wake with a jolt, stagger downstairs, drink a cup of strong coffee and attempt to decipher the powerful dream dynamics. I always reached the same conclusion. It was time for me to return to Wakani and assist a friend in need. Antonio was a brother in spirit and had risked his life on numerous occasions in my quest for Mejia's emeralds. It was payback time.

I called Adriano and told him I would arrive the first week of December. Brit made certain I had meals that didn't require refrigeration or anything beyond heat to make them enjoyable. The two shopped for gifts for the Vintimilla family and paid special attention to things Adriano Jr. and his sister, Maria, might enjoy.

Brett watched as I carefully wrapped foil and Christmas paper around boxes of ammunition. He finally asked how I'd get out of an Ecuadorian jail if I got caught. The thought hadn't crossed my mind and I had a sudden desire to cancel the trip. I verbally rationalized with them that I might be gone over the holidays and maybe not return until the after the first of the year. Both of them said it was something I had to do and they weren't about to let me escape what they called 'my destiny'.

• • •

The muscles in my body tensed as new electronics in the Los Angeles airport scanned the contents of my duffle bags. Hidden among my books and videotapes were a stripped down 12-guage Mossberg and 35 rounds of ammo, wrapped in aluminum foil and concealed in Christmas gifts. Gone were the days of openly packing weapons wrapped in underwear and jeans. I held my breath when the conveyer belt backed up and the scanner took a second look at the contents of my duffle bag. He leaned forward and studied the screen for the longest time. I held my breath. Then in a split second, he seemed to lose interest and passed my bag through the security check.

It was somewhat ironic that some 20 years earlier I had left this same port of call, a wild-eyed kid in search of El Dorado. I thought about how much of a departure this was from back then. I was older and hopefully wiser, however, my feelings about returning to Ecuador were mixed. The anticipated joy of seeing old friends revved my emotions momentarily but they quickly subsided with the apprehension of what might be lying in wait in the Oriente. I had to remind myself that I had no plans for an expedition; I was only on a mission of mercy, yet my instincts told me to return with adequate firepower.

The Ecuatoriana 707 circled over Quito and the visible change in the topography of life spread out below was astonishing. Progress had assaulted Atahualpa's birthplace, the northern fortress of the old Inca Empire, in a mad rush to modernize. The once quaint, old city had dashed headlong into the twentieth century and the most significant difference was in the Quito airport.

Customs had been completely modernized, which was evidenced by closed circuit video cameras that scanned every move. I took a deep breath and approached the customs lady that was poised to search my luggage. Old habits surfaced and attitude became everything. I threw my duffle bag on the counter and ritualistically fumbled for the key to unlock it. I apologized for their misplacement and she waved me through. Her lack of patience was my savior.

I chose not to spend any time in Quito and boarded a SAN 727 packed with well-dressed businessmen and women traveling to Cuenca. In days passed Levis and leather were the vogue but three-piece suits and smart looking ties had replaced that attire. My window seat allowed a panoramic view of lush green as we steadily climbed in elevation. Before long the towering, snow-capped Cotopaxi volcano came into view and played hide and seek before eventually disappearing behind a veil of white puffy clouds. Off in the distance the fiery Sangay volcano, the probable birthplace of Mejia's emeralds, defiantly belched ash and smoke.

Approaching Cuenca I felt a tingly sensation and swallowed hard as we skimmed the red tile rooftops and passed the tire factory before dropping onto the short runway nestled in the city. The old charms of the terminal had been replaced by airport screeners and ever present electronics monitoring weary travelers. Nothing was the same except the familiar faces of the Vintimilla family that seemed more mature, but nonetheless recognizable.

The ambiance of Cuenca immediately recaptured my heart. The city had changed, modernized and grown, but its heartfire was still the same. Adriano was thrilled that I planned to journey to Wakani. He understood the significance of my return. Although aware of the spirit involvement in the Kionade, he embraced my decision. After all, it was his old family home that had indoctrinated me into the supernatural world of ghosts, goblins and things that go bump in the night. Once he knew that I was serious he shared his recent findings on the Mejia project.

"I wrote you about my Mother's death," he said sadly. I nodded in reverence. "What I didn't tell you because I felt it was improper to discuss at the time, was before my Mother died she reverted back to her childhood memories and we discovered many wonderful and beautiful things about her family, the Ordonez. I knew they were wealthy and had extensive land holdings but had no idea that at one time they were the wealth of this country. The Ordonez family had three brothers, one that lived in Paris, one in Guayaquil and my Mother's father, who lived in Cuenca. They rotated their responsibilities that included a holding company in Ecuador and one in Europe. They became quite wealthy through quinine. Surprisingly, I discovered my family was the major stockholder in the company Mejia worked for.

"After you left, I discovered that my relatives financed the cascarilla operation in the Macas Forest. The workers were paid with coins that bore the mark of the Ordonez family."

He took a moment to study my reaction, which was total shock. What a karmic circle we were immersed in, I quietly thought.

"Now what's interesting is that the cascarillas began their operation at Paute on a farm owned by my mother's father and from there they pushed east into the mountains and jungles towards Mendez and north-east into the Macas forest. Records are scarce, but it appears that several years passed before they finally reached the Rio Tutanangosa and placed their base camp at Tambo Santa Rosa."

Needless to say I was speechless and understood his reluctance to let go of a project that had created so many twists and turns in our lives.

He smiled and continued, "Somehow I have confidence in the Mejia project I didn't have before. To know my family started the cascarilla operation in the Oriente and that we are the ones destined to complete this circle is fascinating to me. Lee, we must take another expedition to find the emeralds. "I realize that when you returned from the Kionade, you had lost interest in pursuing the emeralds and your life had changed directions, but now with your dreams of Necta's problems I believe we should give it one more chance. What is the word you used when the timing of unusual opportunities come about?"

"Synchronicity," I replied.

"Yes, that's the word and the reason we should not give up. Those strange coincidences continued to surface. I accidentally ran into a man that claimed to be the great grandson of Raphael Mejia. His name is Segundo Mejia and I invited him to come to my office at the bank to share information on his legendary great grandfather. Segundo came to my office with a friend named Cambesaca who had served with him in the Ecuadorian military. They had searched for the lost mine together. At first they didn't want to tell me anything. Then I quoted word for word from the last will and testament written by his great grandfather. Segundo was surprised that I had this information and began to open up.

"He said that when he was very young his great grandfather (Raphael Mejia) would call him to his bedside each evening after dinner to participate in another lesson. Segundo had to repeat all the details of how to reach the emeralds. Every detail was repeated over and over. Where the rock open in the middle was. What was the color of the stones? Describe the cliffs. Segundo said he had forgotten the details to the cascabel and the exact location of the Tambo Santa Rosa. He said if we helped him find these two locations then the emeralds were easy to find. He knows exactly where to go.

"I told him that a Shuara friend and an associate knew the location of the first Tambo Santa Rosa and offered to put up the money for him to search for the mine with us. I also explained that we were quite certain the cascabel was a series of waterfalls in the river the Colombians named the Santa Rosa. However, I was careful not to mention the true name of this river until after we entered into an agreement. I then offered to draw up contracts for an equal share in the find since we both had separate puzzle pieces for our common goal. The men left to think about my offer and said they would contact me within the week."

His head dropped, "I never heard back from them."

"Here we go again," I grumbled.

"I lost contact with Segundo until one day when reading the Mercurio newspaper searching for investment opportunities I came across a gas station in Gualaquiza that was for sale by Segundo Mejia. I contacted him by telephone and that is when he confessed. He said that his friend, Cambesaca was suspicious of wealthy people. He felt they would be cheated and became adamant they explore alone. At first I was angry but kept my feelings to myself."

Adriano paused and shook his head in disbelief, then added. "Maybe the Alfredo fiasco had something to do with my decision not to pursue his trust. I allowed him to drift away to continue his search with his distrustful partner and to this day I have not heard from him."

Adriano seemed content with his decision. "Like truth, the cascabel was illusive because everyone in his camp had their own version of what it was," he said sadly. "Somehow Segundo's memory has failed him with the important keys to the family treasure because he's still convinced the cascabel was a hill named because of its snake population. The same could be said about the Tambo Santa Rosas. He was confused over which Tambo his great grandfather worked from. He told me once that he had lost interest because he had been told many stories about where the treasure was...It was a puzzle with distorted clues."

Later Adriano heard through other sources that Segundo's enemy was not a distorted puzzle. It was the tigers. On his last expedition a tiger attacked their camp and took one of their dogs. The attack was so lightening fast that they didn't have time to react. The following day they searched for the waterfall described by his great grandfather where the Shuara used to bathe. Suddenly five tigres came off a loma near the Rio Tigress and headed toward them. They had to run for their lives. They watched in horror as the cats began sniffing for their tracks but lost them at the rivers edge, which saved them. This incident frightened him as much as a priest in Sucua that offered to pray for Segundo's soul when said he was going into the Macas forest. Segundo had told Adriano's source that he would die if he went back.

Adriano shook his head in disbelief. "Isn't it amazing that Segundo and Cambesaca don't trust anyone to go with them yet they won't return because they are afraid of the jaguar?"

I understood that fear. Anyone who has heard the low guttural cough of a big cat doesn't forget it.

"Lee, we are kindred brothers when it comes to understanding the psychology of greed mixed with fear."

We both offered a weak laugh with that assessment. Then he asked if I was up for another expedition into the Macas Forest. I nodded weakly. Once more the loyalty issue gnawed at my consciousness. Then I rationalized that since I was going there anyway, I might as well go prepared.

Adriano's brief meeting with Segundo revealed bits of information that shed light on several mysteries. The tigers were still hunting in packs and this led me to believe that our old nemesis, Singusha, was still in command of the night stalkers. I wondered if I could once again muster up the courage to place myself in harms way with these uncertainties?

Another question answered by Segundo was how Mejia moved the 75 pounds of dead weight. Adriano discovered that Raphael Mejia used his coat to wrap around the tres arrobas (seventy-five pounds) and tied it off with his belt in order to drag the emerald laden rock to its final resting place. He hid the emeralds in the large white rock open in the middle and covered them with blue stones.

Raphael Mejia was 90 years old when he died and up until the time he died he believed that his family would find the treasure, but that never happened. He had failed to take into consideration that cascabel had many different meanings in the Oriente and he had no idea there was another Tambo, cascabel and Santa Rosa River.

I was safely ensconced in Maria's bedroom of the new Vintimilla family home and enjoying the company of my old friends, but after a week without any reply from my telegrams to Necta, I was getting restless. Adriano had a multitude of business commitments and was hesitant about leaving for Sucua.

I knew it was time to make my move and find my Shuara friend in the Oriente. Like always the ritual began with packing for an expedition. Adriano Jr. and his father pulled my tents and sleeping bags from their storage place and we checked to see what shape they were in. After all these years most everything was still usable. Planning the journey brought back so many memories but this time around it also took on a whole new meaning.

My values, what the Shuara referred to as a person's heartfire, had changed. I wondered if perhaps it had been lost in the fog of civilization's stern requirements. I always felt my intentions had been honorable, but this trip was straight from the heart and directed by a man who had beckoned me through my dreams. The unta shaman could read the invisible aura of a person's true intentions and interpret the purpose of any colonial entering their land. Necta had referred to one's intent and ethics as heartfire. I knew firsthand that one had to pass their tests of purity of thought and deed before safe passage was allowed. I wondered what difference the unta would see in my heartfire. Here I was in the sunrise of another journey, one spun from a dream, unlike anything I could have ever imagined.

The road to Sucua had stubbornly avoided progress. It was still one large pothole after another and crumbling ground around old wooden bridges stressed by time and decay. The arduous, time-consuming drive caused Ted, Adriano's brother, to wince in pain with each jolt. An operation had gone awry and, still bandaged, he was suffering from the pounding we were taking on what passed as a road. Ted was a kind and caring man and when Adriano informed him that he couldn't make the trip, Ted volunteered his services despite his physical condition.

Time passed quickly on our final leg of the journey and soon we rolled into the settlement of Sucua. It was like entering the past with the exception of the newly cobble stoned streets in the main part of the square. Sucua still lacked development and evolution and was teeming with angry young Shuara who seemed even more distrustful of the colonials of today than in the past. I recognized the hatred in their eyes. If Necta were indeed in danger then it would be wise for us to find a pension (hotel) on the outskirts of Sucua to avoid bringing unwarranted attention to ourselves.

The following morning Ted dropped me off early at the Necta household outside Sucua and drove to Macas in search of something to ease his pain. We agreed to meet later that day. Esus, Antonio's brother-in-law, greeted me excitedly and said that Antonio had been expecting me and would return in one hour. Hopefully, one hour was the measurement of time in my vocabulary, not the Shuara. Esus, to outsiders was still a frightening presence, but to a proven friend he was as mellow as a pussycat. We retreated into the darkened house that lacked fresh air and sunlight and sat cross-legged on the dirt floor speaking in soft tones. I learned the truth concerning Antonio Necta.

"He has lost his heartfire; his eyes offer no life, they are hollow with sadness and anger," Esus said sadly. "After he lost Muriyima (his wife) some members of our family believed Antonio had changed and took control of Wakani. My wife and I share this remaining house with him and try to keep his life force content. He told me that you would return. Senor, Lee you must help him. Only you can."

It was hard to believe that Don Antonio Necta, the pener unta uwisin, was a shell of the person I once knew.

Esus related that there were other problems, too. Antonio was being watched. There were elders within the unta Shuara that believed Antonio had turned his back on the tribe and divulged tribal secrets to the colonials. He didn't say so, but I felt he was referring to our search for Mejia's treasure. This disturbed me greatly. Had the tribal elders mislaid their specter souls for the coveted namura, the green fire? Or were they truly upholding tribal traditions? Had Don Antonio become a marked man because of our friendship and past expeditions?

I knew I had to defuse the doubt around him but I wasn't certain how to go about that. The only thing that came to mind was another expedition together. Then those suspicious of our journey might be convinced we were still searching and no secrets had been divulged. If I could convince Necta to take me into the Kionade my plan might just work.

Within the hour, Necta's high-pitch voice displayed true excitement and emotion when he entered the hut. Joy returned to his weathered face and he really didn't seem to have aged. Perhaps he was broken within, but if Esus hadn't told me I would never have guessed there was a change in Necta. Once inside his primitive domain he reverted into his alpha male status. He was the leader of his pack and a missing member had returned home. Traditions were difficult to break when in the comfort of familiarity and trust. When I asked if he was ready to take me into the Kionade his eyes sparkled and his voice rose in pitch and then he disappeared behind a curtain leading to his bedroom. He returned and unfolded a map. He fumbled to read his handwriting in the dim light. His eyesight was failing. I reached into my pocket and handed him my glasses. The uwisin shaman had become mortal.

His new map was confusing to say the least. The last river in sequence on the north bank of the Tungus was the familiar Rio Ojal. This was correct but below was the Cascabel River instead of the Santa Rosa followed by the Rio Imuirtni. His Santa Rosa was now the fourth stream down river from the headwaters. Did he rename those rivers to throw off those watching him and prowling through his personal items? Was he sly as a fox or suffering memory loss?

He must have read my mind. "Some in my tribe believe I have found the Colombian's green stones in this river, so we will go there and prove those following us wrong." His finger pointed to his bogus Rio Santa Rosa which, in actuality, was the Rio Tigres or what the Shuara referred to as the Huahuayme.

"After they grow tired of our unsuccessful search they will leave us alone. When they leave we will cross the Tungus upstream near the Rio Saladeros." His whispers were interrupted by the sound of Ted's SUV on the dirt trail.

I asked if he still had the Remington shotgun and he replied that someone stole it. I understood my urge to bring in the Mossberg. In a soft voice he whispered that many things had changed. Permission was necessary from the Shuara Federation to go to the rivers. They no longer allowed exploration on their land but would give permission to hunt or to film. To my surprise he had already plotted out the expedition. He would find horses and cargo carriers that could be trusted, and we could leave in the morning.

I had a bad feeling about this expedition. My dreams had led me back to hear Adriano's passion about our destiny to find the green fire that had somehow morphed into a mission of deception to throw off Antonio's watchers. I tried to conceal my concern, especially around Ted. I walked Ted to the SUV when suddenly Antonio called Ted back for some last words. When Ted joined me outside he appeared concerned but nothing was said. I assumed his side was bothering him.

On the drive to Macas, Ted let loose. He initiated a gentle scolding and suggested I cancel the expedition and return with him to Cuenca. When that didn't work he shifted his approach. Directly and firmly he demanded, "Lee, are you crazy?" He repeated the question for effect. "You have one shotgun, a 22 rifle and a guide that has lost his self-confidence."

My quizzical look led Ted to open up. "Lee, before we left Antonio asked me what he should do with your body in the event of your death. I know your friendship with Antonio goes back a long time, but he's changed. He's lost his ability to lead and protect you."

Ted had my attention. I was concerned, but something inside, that little voice from years ago that protected me in the face of danger, was pushing me, urging me to live up to my commitment to help a friend.

When I explained my plan, Ted relaxed a little. "So, finding the emeralds...it's a decoy?" he asked, smiling. "But if Antonio is being watched then you must be careful my friend. There are still the tigers and the snakes and you don't have dependable people. If you want, I will stay and wait for your return."

I dismissed the idea saying he might have a long wait. It fell on deaf ears. "I'll wait until hell freezes over for a friend." He too, suffered from the commitment syndrome. Our mutual stubbornness and strength of commitment created a unique bond of friendship.

That night I had a warning. It came through dream state when I found myself high in the clouds spiraling out of control like a crippled airplane plummeting towards earth. The dream was neither esthetic nor wispy. It was prophetic with real time sensations including reverberating jungle nightlife sounds. I was frozen in fear and couldn't move. I lay in my bed soaked in sweat. It dawned on me that I could change the outcome. For the first time in my life I utilized a technique to change the outcome of that dream.

If dreams were a creation of the subconscious mind and responded to thought and thought could change a given situation, wouldn't it be great if we could do likewise with the conscious mind? Perhaps, I had. Wasn't I the guy that had that protective light around him all those years in the wilds of Ecuador? And hadn't Antonio school me in the past on how to balance fear when danger was apparent? Our survival depended on the power of a positive attitude. I realized that in my rush for closure I had mislaid my spiritual balance because of the uncertainty of my journey. How stupid was I?

It was obvious that I needed shamanistic balance for this journey and protection would be provided through the equilibrium of maintaining my self-discipline. I knew I would be fine if I stayed in tune with nature and those around me. Taoism teaches one to surrender to the Greater Mind that is nature itself. To do this first required finding the will to allow my mental and spiritual balance to come forward. My hiatus from adventure and years submerged in material civilization had subverted my bond with nature. I wondered what had subverted Antonio's?

After a sleepless night we arrived back at Antonio's hut. There were no horses to be seen and I wondered if our trip had been cancelled. Two surly looking men stood near the entrance talking in hushed tones. They looked mean and edgy. When Esus and Antonio came out to greet us, I asked where the horses were. Esus responded excitedly. "Oh, Senor Lee, you will not need horses because there is a new road all the way to the new bridge below Wakani. You can drive there in your friend's car. The federation built a bridge across the Tungus to open new settlements on the south bank."

Ted seemed relieved and joked that they should name the bridge "Lee's Ferry" as it surely contained the steel cable I'd brought to Sucua so many years ago. I laughed but was curious and bothered by why Antonio had failed to mention that earlier. Why did he say that we would need horses? Why was he taking me up the north bank to a fictitious Santa Rosa River? Why did he suggest we cross the Tungus in supposedly shallow water when we could simply cross by bridge? And what about these two sullen men selected to be our cargo carriers? In the past Antonio would never have hired these types. Why now? Who were they?

Why had Antonio made that horrible comment to Ted about what to do with my dead body? I silently admonished myself to keep the balance; not an easy task with two suspicious cargo carriers, an uwisin that had lost his power and possibly his memory and self-confidence. Negative questions replaced my positive thoughts and I remained silent.

The jungle beyond Wakani was unchanged, dense, canopied color meshed between greenish black and brown and with an occasional bonus of color thrown in from a white, red or orange orchid to liven things up a bit. Quick as a cat, Necta unleashed his machete and in a split-second cut one of the beauties and wrapped it in a broad leaf for safety. It was good to see flashes of the old Necta out here in his private world, free of worry and intrigues. Beautiful memories from our past expeditions danced briefly before me then vanished when we pushed ourselves deeper into thick foliage on a direct course due west.

The semi-trail was sporadic and difficult to maneuver. My heavy pack wore down my stamina and turned my legs to jelly. We followed a narrow furrow cut into the steep hill from water runoff and reached the bottom only to wade across another nameless slow flowing stream. The moss covered, round stones polished smooth by the endless flow of bacteria infested water demanded caution. We carefully navigated an accident in waiting only to begin another climb; a scene repeated countless times in the womb of the forest.

It was a struggle to stay upright and to keep from falling a hundred feet when skirting the edge of a small canyon.

My normally quick reflexes turned into slow motion when the heavy weight from my backpack shifted and I lost my balance. My first reaction was to grab anything nearby and, as in the past, whatever I reached for was green or brown. As with all of the previous slips in years gone by, I hoped it was not a hanging snake or rotten limb.

About the time we worked ourselves free of ridge climbing we began another exercise more terrible than the previous. Before us, a thick, head-high palma grove, the natural habitat for palm vipers and four legged predators blocked our path. About the time I was having second thoughts, Necta plowed into the twisted maze with the fervor of a man on a mission. I followed on his heels and could only hope that trouble wasn't on the way. Not wanting to be left behind the reticent cargo carriers fell into line behind me. Moments like this, when the blind spots in the jungle distort rationalization and shake your faith made me question my sanity. My focus on finding a balance with nature vanished and was replaced by the safety of numbers and surprise.

The scattered rays of sunlight told us when to build camp and we obeyed its visual command. I was tired, angry and wanted to rebel against the punishment I was putting myself through. I hadn't prepared for a serious expedition and I felt like I was in a crash course of bodybuilding. We were short one carrier. I'd been lugging fifty pounds of backpack crammed with food and extra ammo that left me exhausted, which was obvious to everyone.

My hands trembled when I offered the cargo carriers American cigarettes. They took advantage of my discomfort, threw their heavy loads on the ground and complained about the weight and poor money they were making. They wanted six dollars a day per man or else they were going home. Necta ignored their warning and wandered away seemingly lost in thought. We'd only been on the trail for six hours and there was already mutiny in our ranks. Loyalty was in short supply these days.

Too tired to argue and in no mood to cook for the crew, I fished cans of assorted food and crackers from my pack. That triggered another round of complaints and menacing stares. Their anger exploded when they discovered they had to eat from latas, (cans of cold food). They informed us they were leaving at first light and insisted on being paid six dollars per man for their work. I quietly paid the men in sucres, but instead they insisted on being paid in dollars. Good riddance. I slapped ten dollars on each open palm and told them at first light to take our extra packs to Wakani and leave them. They grudgingly accepted my offer and disappeared into the two-man tent I'd provided. It was the beginning of a long night.

In the old days Necta and I would spend our evenings talking about many wondrous things until exhaustion demanded sleep. We slept peacefully back then because our men were handpicked for their loyalty to Necta and trustworthiness. But, those days were gone.

Unconcerned, my friend fell into a deep sleep. With my bed companion, the Mossberg, snuggled against my side I waited for an outburst from the two whispering warriors but soon the conversation from the other tent grew muted. The cargo carriers went to sleep. I, too, soon joined them.

The Shuara cargo carriers left us at morning light. Necta seemed content that we were now alone. He helped me strike camp and insisted we leave immediately to find shallow water to cross the Tungus into the Kionade. He was adamant that those following us would turn back because of their fear of the Kionade. I asked who might be following us and he said those that had been watching him for years. He implied that the men that left had no stomach for work and were spies for the federation who would wait on the trail to see if we had the green stones when we returned. Unfortunately, my plan to bring him back into the fold with a fully staffed expedition was unraveling quicker than his mood swings. He seemed out of control.

The next day we found ourselves balancing precariously on a moss covered log attempting to cross a lagoon. Suddenly Necta slipped and started to fall into the water. I grabbed his pack and righted him but then I slipped and fell backward into the pool of still water.

My head struck a rock and I briefly lapsed into semi-consciousness before the cold water revived me in time to see a long thick shape coming toward me. It stopped a few feet away. I saw the thick rope-like object had a snake shaped head and fluid black eyes. It was golden in color and stared curiously as if to study me. Strangely, I didn't feel fear, only curiosity. I knew I must rise to the surface for air but I didn't want to. I wanted to stay and admire the beautiful creature.

When I popped to the surface gasping for air there stood Necta on the bank with a blank look on his face. He reached down to give me a hand and I noticed he had no emotion. There was no joy, no fear, nothing except a distant stare. I excitedly asked if he saw it. There was no response. When I scanned the reflective pool there was no sign of the boa, only elongated ripples of disturbance.

As we walked on I couldn't shake my boa experience. It was too real. Had I encountered the legendary shaman, Tsuni, the shape shifter that could take on the appearance as a golden boa? Was he giving me a final warning to turn back? He hadn't been aggressive, so maybe there was another message, one I had yet to comprehend? Necta's indifference to my close call and boa experience hurt as much as my aching head. The warning had been swift and painful. It was time to call a halt to this bewildering journey and Necta offered no resistance.

There was more strangeness waiting when we finally arrived at the edge of Wakani. The sorrowful description by Esus of a deceptive family that stole Necta's land had prepared me for the worst. But, instead of a doom and gloom environment there was joy and happiness at Wakani. Women, old and young, in the spirit of cooperation attended to a large garden abundant in fruit and vegetables. Along the river's edge others washed themselves and small children who chatted and laughed with each other while laundry dried in the sun on large rocks. Men were serenely chopping firewood and performing other tasks around the farm. Children playfully chased each other in a field of green grass that was dense jungle before Necta and his vaqueros cleared the land. The only thing missing from this peaceful scene were unicorns cavorting beneath a multi-colored rainbow.

The soul of the land pulsed with the joy of life, which made me wonder if I had died and gone to Shuara heaven. The painful lump on the back of my head told me otherwise. When the children spotted us they began yelling in their native dialect and ran toward us. Several women also approached us and showed their excitement and joy when they saw Necta. A middle-aged woman that I later discovered was one of his many relatives, called out to come and join them. When he balked, the children took his hand and led him to the compound. I was taken to my backpacks that the carriers had dropped off and then told to sit and rest while warm food was prepared. The women surrounded Antonio Necta offering to attend to his needs. He seemed embarrassed and woefully detached to the sentimentality and respect shown to him.

Necta's stubborn reticence appeared to have underlying foreign tones to his family drama. That evening the new matriarch of the Necta clan along with others called me aside and thanked me for returning to help Antonio through difficult times. I couldn't remember meeting these people but that was unimportant because they definitely knew everything about me. Apparently my exciting adventures with Antonio Necta had been handed down through the Necta clan. It brought back memories of the adulation and respect I received years ago on the wind swept plains of Nabon.

My hyper-cautious defensive mindset heightened when they repeatedly told me that Antonio had changed. I had seen it firsthand but asked them to be more specific. What had caused the change? I was only told that he had forgone tribal traditions in favor of the colonials' way of life and because of this certain members of the tribe were watching his activities. I asked if I was responsible for him being ostracized from his tribe?

In unison they responded with an adamant "No! He is in danger and it is not because of you or what you search for. It is much bigger and only concerns tribal traditions. He knows what it is he must do but he refuses."

I wondered what the root of the trouble really was, but understood they had shared what was permissible.

When the fire had died down that evening and the household was still, I wrestled with the day's events and how to safely return Necta to his uwisin status? It would have helped me immensely if I had had an inkling of his problems. My questions seemed to magnify the pain throbbing in the back of my head. Finally, after gulping down extra-strength aspirin, I fell asleep.

Once again a dream played through my mind. It was realistic and contained the reverberating sound of rushing water and a cable strung basket in the foreground. The interesting aspect was the emergence of day and night within the same scene. In dream state I realized I was being given a choice. I could enter the darkness of despair, turn my back on a friend and walk past the bridge down the road towards Sucua or position myself in the basket container and pull myself across the river into the sunlight of discovery. The dream gave me a choice. Without hesitance I made my decision to cross the Tungus and allow fate to be my accomplice in the Kionade.

The next morning Antonio and I, along with two youthful volunteers from the Necta clan, crossed over the Tungus by cable and followed a trail along the left bank of the Cunquinza. This was the same area I had explored years earlier but since then it had changed dramatically. A well-used trail by those that had settled in the area had eliminated hours of chopping through dense jungle. Many trees had been cut for construction of new houses and cleared for pastureland. Progress had arrived below the Rio Cansar, but across the Cunquinza the old Kionade was still unchanged by time or progress. It was as dense and foreboding a jungle as before without any signs of trails, houses or movement. It still carried an ominous presence and apparently was off limits even to Shuara

Federation land grants.

Late afternoon shadows started to creep across our trail; Necta snapped out of his previous role as silent observer and suddenly displayed the leadership qualities of old. He called a halt to our undefined journey and selected a campsite in a dense area of thick foliage with large trees near a rocky ledge close to the Rio Cunquinza.

In a burst of newfound energy he took the lead and began cutting a place for our tents totally unaware that his youthful admiration society watched every move. When we were in our prime blazing new trails out here, these boys were too young to participate and could only partake in our adventures through stories that left them dreaming that some day it would be their turn to master the jungle. This was that day and they studied every swing of our machetes. They were blazing a trail with Don Antonio, the tribal uwisin that defeated waweks and cured evil through magic and nature. That was the real Don Antonio, not the one whispered about or the one that some said had betrayed his people. If only Don Antonio could recognize this and rekindle his heartfire.

That evening while Antonio stoked the roaring fire and tested his strong coffee, I plopped four ready made meals of rice, veggies and sweet and sour chicken in a boiling pot and within minutes my white man magic had won over the hearts and minds of my crew. Thanks to good food, youthful enthusiasm coupled with respect for their elders, and Necta's new peace of mind, we were off to a great start. The balance of harmony had returned to our expedition. Now if it could only find the heart and mind of Antonio Necta.

The boys retired early to experience their little, blue, domed tent with windows and their animated excitement drowned out the rushing water of the river. While they chattered like crazed parrots and giggled like the kids they were, Necta stood near the riverbank and gazed across into the Kionade lost in thought. I leaned forward and removed a burning stick from the fire, lit a cigarette and leaned back against my pack. This was unlike me; normally I would have been in the safety of my tent by now, but not tonight. I wanted to feel the nightly pulse of the rain forest and for the first time in my series of journeys I felt in control of my destiny.

Night sounds from the jungle seemed to calm rather than warn. I felt in unison with the lowering and rising crescendo of the pulse of the magnificent forest around me. Necta sensed my balance and joined me. At first, he said nothing and we both stared aimlessly at the crackling fire. Then in a soft voice, almost in a child-like manner, he asked me a surprising question.

"Lee, do you believe in the Bible?"

"Yes," I replied curiously.

"Do you believe that those who accept Christianity will have a mansion waiting for them after they die?"

I replied that I believed that people would find what they wanted when they died.

"Me, too," he said smiling. He then stood up and patted me on the shoulder, thanked me for being a friend and went to bed.

A tear slid down my cheek. My grandmother and her Pentecostal faith came to mind. The devout believed a mansion on a street paved with gold awaited them in the afterlife. This was of great comfort for her in those final days, as I knew it would be for Antonio Necta.

My uwisin friend had been seduced by promises of things to which no one had a definitive answer. I could only imagine what was going through his head. I realized my friend had gone through years of debate with himself but now seemed content to believe that his conversion to Christianity, as well as his final days on this earth had purpose and meaning. Antonio mistakenly believed he had lost his beloved Wakani to the non-believers but now a mansion awaiting his arrival would alleviate this loss. His struggles down that long winding trail in search of the truth between his surname Necta and Christian name Antonio were over.

I sat in silence. Thinking about what he had said. Then I remembered when I first met him. The signals were apparent back then when he confided how his decision to combine the forces of nature with the power of the supernatural went against the grain of pure tribal shamanism and forced him to walk a fine line in the Shuara community that he served. In order to appear different than the normal uwisin, he adopted western clothing and did not confront the colonial's religious leanings. In doing so, my old friend had turned his back on the age-old traditions of his tribe in order to accept Christianity. What possibly could have turned him against his deep-rooted Shuara shamanism? He had always fostered distrust of the black-robed teachers and the Church they represented. His conversion to Christianity was acceptable but the danger for doing so was what made my mind spin. That made no sense at all because many Shuara were accepting the white man's religion.

I sat still...thinking. Was he facing his mortality? Was his body getting old and tired from its struggle in the primitive surroundings? Perhaps he was ready to leave the Earth and had a newfound courage and understanding that replaced his Shuara teachings. I understood why my dreams led me to this moment in time. He respected me as a friend and someone he trusted. His confirmation of his belief came from someone who would not lie to him. I always gave Antonio the truth, my truth. As with life, I believed that death is one's dream or vision quest of what each individual perceives as truth. My grandparents believed, as Antonio, that they would go to the Promised Land and live comfortably. Who was I to argue with their perspective of truth?

When Antonio first entered my dream state and beckoned me to join him, I didn't understand why. I had misinterpreted the gold held in one hand and a green stone in the other as a materialistic gesture of discovery. I now knew that was not the case. He had not found an outcropping from the tears of the sun or the green fire of Mejia. The gold was symbolic of the Church and its adornment of the metal. The green stone was the sacred namura of the Shuara tribe, the emerald and the color of the heart chakra. In my dream Antonio was holding both and appeared confused whether to choose the left-handed path or the right-handed path? My destiny with Antonio Necta was to hear and accept his new beliefs, which I did and for that he was grateful. My mission was fulfilled and I felt relieved. The smoke from the campfire burned my eyes and nostrils; I decided it was time to turn in.

CHAPTER 19

IWANCI

(left to right) Lee Elders, Antonio Necta , Marcello Natema

The following morning my peaceful sleep was shattered by an angry conversation outside the tent. I poked my head out and saw Antonio being berated by a young Shuara, that I guessed was in his early twenties and who was waving a lethal machete. My first reaction was to grab the Mossberg to even the odds, but I laid the weapon down and slowly crawled out to the surprise of the young buck. When our eyes met he immediately switched from Shuara to Spanish and ratcheted down his anger to a more manageable tone. He demanded to know who we were and why we were camped on Shuara land.

As he spoke he repeatedly slammed the machete into a tree trunk. His machismo attitude scared the shit out of our crew. Antonio stood motionless, arms hanging limp and head bowed while the boys cowered in fear in the background. Without breaking stride, I smiled, walked towards him and reached down near the fire and poured a cup of hot steaming coffee. The intruder's eyes followed my every movement and he seemed curious by my unorthodox style that was a mixture of unconcern and bravery. Who was this white man that had no fear of the Shuara? When I handed him the cup of coffee I knew that for a Shuara to refuse a gift was an insult and if accepted he must return the favor. He accepted.

He was tall for a Shuara with chiseled facial features and a powerful presence that reflected ancient wisdom through his black eyes. We studied each other through constant eye contact and searched for some common ground. Our conversation was stilted at first but when he learned I was an American that revered his land and had returned to explore its beauty and wonderment he relaxed his grip on the machete and accepted my truth. We moved to a log and sat down next to each other. To any outsider we would have looked like old friends that had run into each other on the trail. His name was Marcello Natema and his family was custodian of the land known as Kionade across the river from where we were camped.

His reciprocal gift to me was an offer of knowledge blended with concern. He wanted to know why we placed our camp in such a dangerous place. He stared at Antonio, who suddenly looked older than his years. Then he loudly proclaimed that whoever chose this campsite made a terrible choice.

"Look," he said. "See that tree leaning over your house?" He pointed and I stared in disbelief. A dead snag that must have weighed a ton was perilously tilted toward our tents.

"See that rock ledge," he continued. I swiveled my head in the opposite direction. "It is the home for a terrible snake that at night leaves his house in search of prey. Last night the smoke from your campfire saved your life if you were outside your house..." His voice trailed off. "Who did this?" Who placed your house in this terrible place?" he asked angrily.

Marcello saw me glance towards Antonio and reacted. "Old man, you are not fit to lead the American's expedition. Who are you?"

Antonio answered.

"I know you from my father. He said you own the land across from ours and you are an unta, an uwisin with great knowledge. Necta, where has your knowledge gone? You are Shuara and should be ashamed of your actions. You have misplaced your Shuara knowledge. I will lead the American's expedition and you will follow."

Antonio offered no verbal response and lowered his head in obedience. The changing of the guard had happened before my eyes. The torch had been passed from the old to the young and Antonio seemed relieved.

Suddenly, one of the boys screamed, "Equis! Equis!" and pointed to the rocky ledge.

We all turned in time to see a two-meter snake with a thick body and slender head come charging out of the rocks, head riding high, less than thirty feet from where we stood. The speed of the deadly reptile was astonishing. If it had been aggressive our only defense was Marcello and his machete. Fortunately, it had only escape in mind and in a blur it disappeared off the ledge with the boys in hot pursuit. I ran the opposite direction to the tent and grabbed my camera. When I returned the snake was in a coiled defensive position and I took my first photo. I had always wanted to see this legendary snake. My hands were shaking when I snapped the pictures even though I knew the photos would be out of focus.

Suddenly, the equis became aggressive and turned to face us. It was in attack mode with only a five-foot ledge separating us from a nasty situation. Someone threw a stone at the magnificent creature. It found its mark and crushed its skull. The brave equis writhed in pain and coiled and recoiled in a fatal circle until its life force left its body. I felt sorrow for the loss of the creature and the pain it suffered. The boy's laughed and congratulated each other on their marksmanship and bravery. Marcello and Antonio sat quietly by the campfire drinking their coffee, seemingly disconnected from what had just happened.

Then Marcello spoke. "They are young and stupid to the way of the Shuara." He said in a loud voice to admonish their actions.

"The equis did not attack and only tried to escape the smoke from the campfire."

Then he spoke in mystical terms describing an integral group consciousness between the true Shuara and nature's creatures, how his father taught him not to kill for the thrill but only for survival. If a Shuara was stalked or attacked by a tigre or culebra then and only then was it acceptable to take the creature's life. He shook his head and returned to his coffee.

The silence was awkward. To change the subject, I shared my experience of how I took the fall for Antonio and my underwater confrontation with a golden boa. I expected him to nod in approval.

Instead his eyes widened and he screamed, "Iwanci is here, we must leave this place!"

I asked what the problem was, but he didn't answer. He hurriedly instructed the boys to break camp with an emphasis on "peligroso"—danger.

We were fleeing some impending danger that only Marcello understood and that seemed to be the result of my encounter with the golden boa. We marched single file in thick but manageable rain forest close to the Cunquinza River. The trip affected each of us differently. Marcello, the new self-appointed leader of our pack maintained a torrid pace and his strange behavior and actions told me someone or something was dogging our trail. My mind was filled with the common, logical dangers of the jungle, but none made any sense. Once again I found myself led into unknown territory by a Shuara that I barely knew.

Antonio slogged along behind Marcello, nursing his walking stick, oblivious or unconcerned with any danger that lurked nearby. The boys, laboring under the semi-heavy backpacks, seemed more tired than frightened. Marcello abandoned the well-used trail in favor of shallow water from the river. Why? It appeared his purpose was to cover our tracks? Why did my boa story turn the son of a shaman into the son of Tarzan? Was our danger created through the puppet strings of a wawek? Were we running from the wrath of an avenging soul? My mind was numb from speculation and my cradled Mossberg weighed a ton by the time Marcello found a clearing and ordered the boys and Antonio to make camp.

In the past few hours my life had run the gambit from fear and uncertainty into the comfort and the safety of a spiritual experience followed by enlightenment and back to fear and uncertainty. And it had all transpired between breakfast and dinner.

We were camped at the doorway of the two mystical Kionade's and I still didn't know why. Unlike my past journeys when the energy was wild and all encompassing it now felt confused and out of sync. Perhaps the Shuara settlements in this area, although sparse, had somehow diffused its heavy influence. Across the Cunquinza, the first Kionade stood majestically unchanged in its defiance of progress. Only one clan lived there and it was the family of Marcello Natema, my newfound friend and protector.

As darkness descended our fire cast eerie shadows that danced around the tree line. Although dead tired I huddled around its warmth and removed my boots to massage life back into my cold, numb feet. The boys were questioning the wisdom of volunteering for our expedition and they weren't alone in their reticence. Antonio appeared content within his silence, a blank stare lost in the fire's hot embers.

Marcello seemed more relaxed, more approachable. I sidled up to him and asked the question that had tormented me for the past six hours. "Iwanci," was the same reply as before. I shook my head to let him know I did not understand the meaning of the Shuara word. He reverted to Spanish. The translation sent chills up my spine. The word he used was demons. His answer created even more confusion. Why were demons after us, I asked quietly? Before he could answer, the stillness of the night was broken by the recognizable coughing sound of a jaguar.

"Tigre," he cried and grabbed his machete and disappeared quickly into the jungle.

My God! This had to be the bravest man I had ever known. To go and confront a jaguar in the blackness of night armed with only a machete was sheer insanity. I sat with my mouth open and stared into the black void expecting any moment to hear the scream of man or predator or both.

It seemed like an eternity before rustling brush announced Marcello's return. He sat down, grabbed his coffee and with a wide grin informed us that the tigre backed down and run away but not before the cat was told that to return meant death. He then became mellow and said he must leave for his village because his young wife was expecting their first child.

Before he left I asked again why he believed we were in danger by demons. "Senor Lee," he said quietly, staring at Antonio, "You and boys not in danger. Its old man them after."

For the next few minutes he wove a mind-boggling story of how an avenging soul, or what he called a muisak, could form into one or the three types of Iwanci, or demons. These three demons could take natural forms and kill the human who sat in their displeasure. One of the natural killer's could be a large tree in the forest that falls on the targeted victim and kills him. I knew he meant Antonio's dangerous camp placement. Another demon or killer could take the form of a dangerous poisonous snake, a makanci. Again, Necta placed our camp near a rocky ledge, the natural habitat of the equis snake. The third natural form of a killer was the boa (pani).

It wasn't until Marcello heard the story of the boa and how I took Necta's fall that he realized we were under the evil spell of a revenging muisak. The tree, the makanci, and the pani were the three traditional forms of death utilized by a muisak. Deaths by these seemingly innocent means were considered accidental in other cultures, but not out here in the mysterious land of the Shuara. According to Marcello these concurrent incidents were premeditated, damning works of magic instigated against Antonio. The disclosure left me feeling that it wasn't the tribal federation that placed Antonio's life in danger, instead, it was an Iwanci sent by a disgruntled wawek. Perhaps it was one of Antonio's old adversaries?

My nerves were still taught when I asked about the jaguar. I knew his answer before the words left his mouth. Still, his answer was chilling. The Iwanci took the form of the jaguar and appeared near his victim to let him know that death was imminent. He did this through the recognizable jaguar cough. That was his signal to the victim. Because of Antonio's new faith he had ignored the warning signs. Or perhaps he had just accepted them as his fate. His Christian faith had led him away from shamanism and had it not been for Marcello the primitive black magic of the Iwanci might have succeeded.

I asked Marcello how he stopped the avenging jaguar soul from carrying out its mission. I expected a macho answer. Instead he informed me that the Iwanci was not afraid of him but his shaman father, the keeper of the secrets. His father evidently had the power to break or remove a curse from an Iwanci. Without further explanation he said he must leave to join his family. I gave him two ready-made meals of sweet and sour chicken with instructions. He thanked me and asked what I wanted in return.

"I want protection for my friend, Antonio Necta," I replied.

He assured me that his father would take care of the problem and return the demons to their sender. With two ready–to-eat meals held high in one hand and a machete in the other he waded across the Cunquinza and disappeared into the rain forest.

After dinner I retired early. It wasn't like I was sleepy, to the contrary, my mind was fried and in need of some quiet time to try and sort out the strange implications of the day's events. Necta, the uwisin and guardian of tribal secrets had become vulnerable when he accepted the white man religion. This allowed his adversaries to attack his weakness and exact revenge. In days of old, he would have reached within his being and consolidated his inner power to fight off the Iwanci until he could journey to the sacred waterfalls and renew his confidence and strength. Once there his vision quest would provide clarity and purpose and his magical darts, the tsentsak, the spiritual servants that resided in the uwisin body would be replenished and power given to destroy or neutralize his enemies. But, that was before he surrendered one power for another.

What a strange drama of human spirit I was immersed in. Necta accepted the truth of his salvation and was content with his decision. But the more he tried to separate from his tribal beliefs, the more events conspired to keep him locked in. His new beliefs precipitated his death sentence initiated by an Iwanci and a forced march by all of us to escape demons. But then, true to the magical timing of this land, a death curse was revealed and terminated by a brash young Shuara who just happened to be the son of an ultra-unta shaman who was the keeper of the tribal secrets. Who could doubt divine intervention after a day like this?

The following morning, after the best sleep I'd had in weeks and two cups of strong coffee, Marcello entered our camp. When our eyes met a warm smile crossed his sculpted face. He carried a burlap sack in one hand and a firmly clenched machete in the other. He excitedly proclaimed that the meals I gave him were the best he and his family had ever eaten and for this he had brought three gifts.

Two were non-physical. They were promises embedded on his family's honor and reputation. He placed his hand on Necta's shoulder and the two men spoke privately for several minutes in their native tongue. Marcello later told me privately that his father had ordered him to apologize to Necta the uwisin for scolding him and to assure him that tribal differences had been resolved. Antonio Necta had been saved twice, first by his savior, Christ, and later by his protector, Natema.

The second gift was also a rewarding promise from his father. Marcello had been instructed to offer me his friendship, trust, and protection during my stay on Shuara land.

The third gift was a burlap sack of sacred ayahuasca. In Quechua, the word aya means 'spirit' or 'ancestor' and the word huasca means 'vine' or 'rope' This was the most precious gift they could offer to an outsider; a dozen sticks some eight-inches in length and dark grey in color from the shaman pharmacist of the Shuara.

The ayahuasca or natema, grown on sacred land in the Kionade, took the name from the family that grew it: Marcello's family. The Shuara utilized a total of ten hallucinogenic drugs and each was taken for a different purpose, but of all, natema, the plant of the spirit, was the most important. Without taking natema one would not understand the others.

Marcello offered to take me to the forbidden land of Kionade. I eagerly accepted and quietly remembered to ask permission of the forest guardians. The first time we crossed into the old Kionade was difficult for me. When Marcello saw my hesitance he laughed and told me not to worry because, in his presence, I was safe. One of my vision quests was reached; I had the freedom to roam the Kionade without fear of reprisal from the unpredictable forest guardians. With Marcello at my side, I was reminded of my youthful exploration in the Arizona desert with the Apache Kid. How unique that my destiny's path for adventure began with an old Apache medicine man and warrior and culminated with a young Shuara whose father was the keeper of secrets of the tribe and custodian of the sacred Kionade, the portal into the legendary City of Magicians.

One afternoon we found ourselves on a game trail that led in the direction of several pyramid shaped mountains covered in thick jungle. One small loma held my attention because of its unusual shape. I unleashed my camera and took a photo of the odd shaped hill with a lone tree sitting on top that gave the appearance of a sultan's hat complete with plume. Was this the place that contained the rock open in the middle that the psychic Richard Ireland spoke of? I searched along the riverbank for a hint of green fire. Chunks of white rocks were everywhere. I picked up some samples and vowed that even if the matrix contained the green fire I couldn't—I wouldn't—divulge their location. Marcello broke more of the white stone from an outcropping, handed them to me and said we must turn around and go back because we were near the City of Magicians. That was the one place we were not allowed to visit.

My eyes strained to see remnants of an ancient city hidden in the jungle. Unfortunately, my physical prism of observation limited its discovery. The City of Magicians was off limits to the novice. Marcello said that in this magical city all past, present and future secrets were known. The shaman vision quest was to gain or earn acceptance to the sacred city where the magic of nature was still present. Not all succeeded, even with the powerful tool of natema that became the trigger for communication between man, nature and Creation.

My quest had brought me to the doorway of the sacred city but in order to enter required a purification process of fasting for four days and an absence of material and negative thoughts. If one succeeded in the mental and physical preparation period then natema could be taken to breach those invisible barriers that protected against misuse of the sacred hallucinogenic properties. However, the key was the maneuverability around and through those psychological hurdles that could trip one up at any given point. But if the initial purification was correctly observed, one could also safely enter this pathway through meditation. This unseen world had many doorways and there was no guarantee the one that led to the city would be opened. Marcello slapped me on the shoulder and laughed. His father said that if I ever decided to walk the path of natema it would require years of thought and deliberation in the comfort and safety of my private world. He knew me well.

Marcello had never been to the City of Magicians either through meditation or natema and emphasized that he had no interest to go there.

"Many years study," he said convincingly.

The length of the vision quest depended on what the vision seeker sought. The quest offered nothing in the realm of the physical, only the spiritual. The patriarch Natema believed that the ayahuasca grown only on their sacred land was the true soul vine that offered the ultimate link to the recipient's higher self and opened the final pathway to communication with the soul. This was the ancient Shuara way.

On the trail back Marcello told me an interesting story about the misuse of natema. When the conquering Spaniards discovered ayahuasca they took the potent drug to learn the secrets of the hidden treasures of the Inca. Instead they became deathly ill and many died. The vine became known as the "rope of death." He laughed and in halting Spanish said that the rope of death was a befitting name from a culture that abused its true purpose. He explained how their preparation process was flawed. Metal rather than earthen pots were used to boil the vine to its resin state, a mistake that proved lethal.

When Marcello entered manhood he took the soul vine to make him strong and fearless and he saw the power and beauty of nature. It was his indoctrination into the spirit world where he met the legendary Tsuni and garden spirits and magnificent holographic images of giant jaguars that he had to touch to prove his manhood.

The Shuara manhood rites lasted for seven days during which time they remained in a spiritual state induced by the soul vine. During this period the young men were taught self-confidence and to believe that the true purpose of each man was to acknowledge they were, through their individual gifts, an integral part of the tribal group consciousness. They were taught how to remove the power of fear and self-doubt from their personal consciousness. Natema was a learning tool that broke down those invisible barriers and provided answers to specifics and the wisdom to understand.

When I asked why so many young men in Sucua appeared angry and distrustful of the colonials his answer was they had severed their connection to the group and were now lost.

"Like Necta," he said. "He's lost but within, he's also content. The young men in village have no contentment, only anger and resentment. They lost attunement with self and nature." He looked me directly in the eye and said, "You...you are more Shuara than some Shuara."

I appreciated the compliment and asked if his father knew of the powerful unta panu to the north to which he answered that his father was a shaman but did not share in knowledge with other shaman because he was the keeper of the knowledge. I shook my head. In the past, Necta had journeyed a great distance to seek out the panu to trade Necta's wisdom for forbidden knowledge. Yet less than a few hours walk from Wakani was the supreme panu of the Shuara and his name was Natema. The Rio Tungus was only the first barrier that protected outsiders' from entrance to the treasure of Mejia and the sacred city, and land where the soul vine was grown. Somewhere in the old Kionade there was a shaman named Natema that zealously guarded those secrets and had the "sight" into the souls of those that came without permission.

I wondered if the City of Magicians was the source of all of the shamanic power in the Oriente? Undoubtedly, this mythical and mystical city was the birthplace of their tribal shamanism and they revered it as such. I remembered that shamanism derived its name from the Tungus people of Siberia and that Tungus was unexplainably the name of the river that ran through the Kionade, separating the north bank from the otherworld. That thought seemed appropriate for my last day in paradise.

As the tents came down I asked Marcello if his family knew of a Colombian named Raphael Mejia. His smile left me perplexed. Then he said that his grandparents spoke of this kind and gentle man that befriended their friend Zamora Singusha. He pulled me away from the others and in whispered tones, told me that his father had anticipated my question. His father knew that I would ask about the Colombian because my vision quest had been to find the sacred namura. He gently added that in the past I had failed in my quest for the namura because my intentions were confused and based on the material, not the spiritual. But that had changed now. I had been allowed permission to enter the Kionade because my heartfire was dedicated to help a friend, a Shuara, find his truth. Because of the change in my intent, the guardians of the sacred land, including Singusha, had accepted me.

• • •

Sadly, it was time for me to return to the civilized side of life. When we arrived at the Rio Tungus Marcello seemed disappointed that our journey was over. I was, too. Before climbing into the bucket to pull myself across, Marcello smiled and said that shaman power comes from heart and namura opens heart to spirit. He knew the deep green color of the precious stone was symbolic of the heartfire of the people—the Shuara. It was their power.

I wondered what might have and could have been with a level playing field. Then I understood what Richard Ireland meant when he said, "My destiny was being played out."

My destiny didn't revolve around plundering Mejia's treasure; it was more complex than that. I had traveled a spiral of expeditions that eventually led to the threshold of my most important discovery: my soul power, my arutam—the discovery of my true self. And Necta had found his savior. What a strange twist of our role reversals.

When we bid our final farewell in Sucua, I gave Necta my reading glasses. He looked puzzled.

"To help you enjoy your new home," I said.

He knew I was referring to that passage in the Bible and smiled. We shook hands and I walked towards the road. I wondered if I would ever see my old friend again? Would he find the balance between his tribal heritage and his new religion? I wondered if Mejia's emeralds would be found and would the Shuara people benefit from the discovery or would they remain forever hidden in the Kionade?

I knew that my friend Adriano would be saddened by my decision to forgo any future expeditions but I felt he would understand my reasons. I too, was saddened, yet somehow relieved that my quest to find the emeralds of Mejia was over. I understood what Marcello meant when he said that the sacred namura was not for one, but for all.

My adventures continued...but not in the womb of the Macas Forest.

Author's Note

After reviewing my reel-to-reel and cassette audiotapes, researching my diaries, examining countless photographs I began writing about my adventures in Ecuador. The end result was a 1200 page manuscript. It was suggested that I divide the book. I did and the end result was two very individualized books.

My early years in Ecuador became *Expeditions: Gold, Shamans and Green Fire*. *Pushin' My Luck*, the second book covers my latter years in Ecuador when my exploration took on a new meaning with the discovery of life-changing, long lost secrets from World War II as well as the adventure of expeditions.

For purposes of clarity and continuity *this* book follows the search for the lost Mejia emeralds from beginning to end. A few of the names of people have been changed to insure their privacy and protection. However, the clues to Mejia's treasure including the names of rivers and the regional villages such as Sucua have not been altered in any form.

For those adventurers that read this book and decide to follow in my footsteps, be aware that time and nature has changed many of those old landmarks and be informed that the Shuara Federation now controls access to and requires permits for exploration of the Macas forest. The days when one could explore freely no longer exist.

Ironically, Sucua is no longer a jungle outpost but a thriving town that has been touched by the 21st century. Perhaps the energy of the forest guardians has also succumbed to progress? Only time and future expeditions will tell. Good luck.

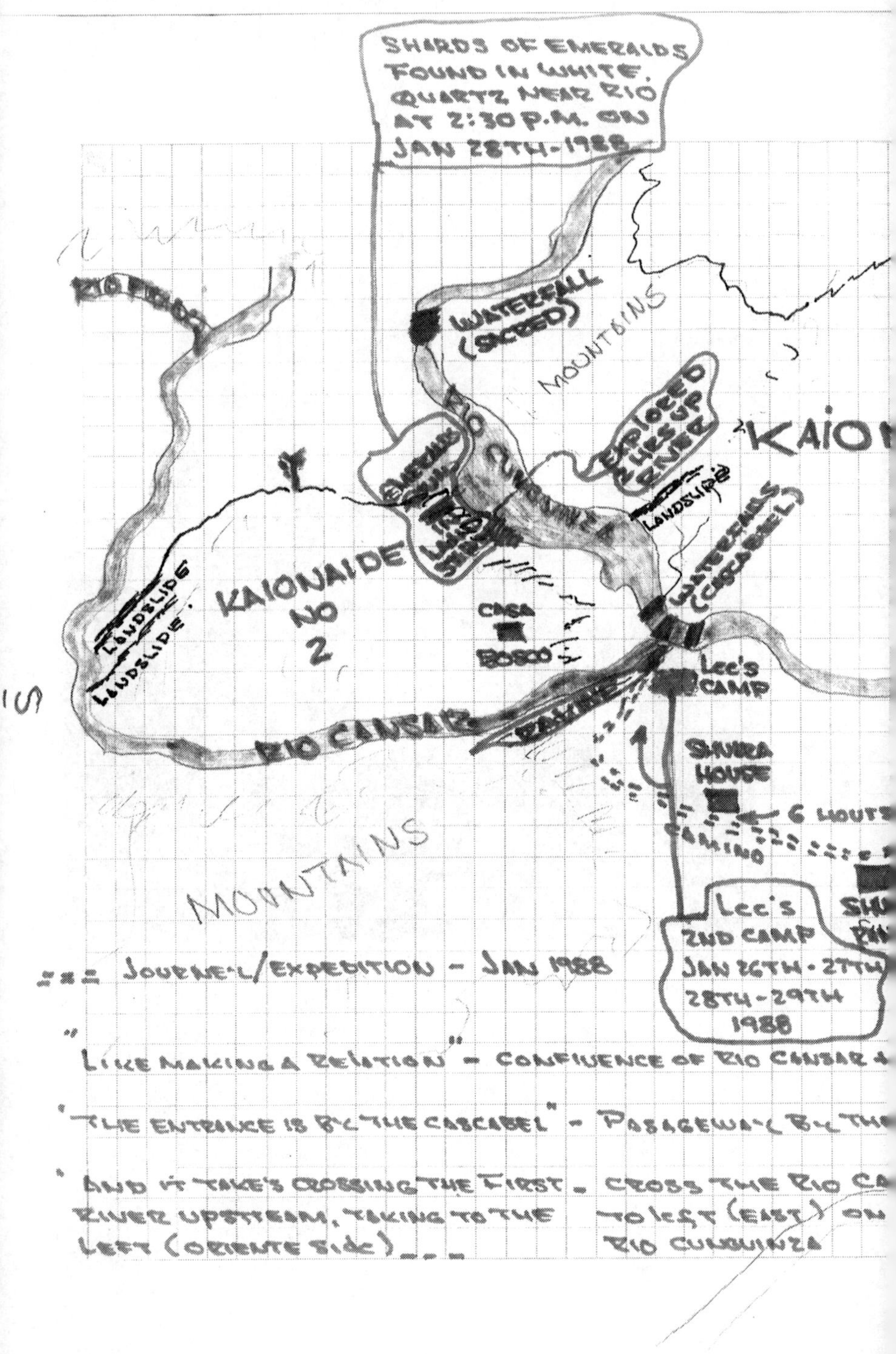
SHARDS OF EMERALDS FOUND IN WHITE QUARTZ NEAR RIO AT 2:30 P.M. ON JAN 28TH-1988
WATERFALL (SACRED)
MOUNTAINS
EXPLORED 2 MILES UP RIVER
KAIO
LANDSLIDE
WATERFALLS (CASCABEL)
KAIONAIDE NO 2
LANDSLIDE
LANDSLIDE
CASA
BOSCO
Lee's CAMP
RIO CANSAR
SHUARA HOUSE
6 HOURS
CAMINO
MOUNTAINS
Lee's 2ND CAMP JAN 26TH - 27TH 28TH - 29TH 1988
= = = JOURNEY/EXPEDITION - JAN 1988
"LIKE MAKING A RELATION" - CONFLUENCE OF RIO CANSAR &
"THE ENTRANCE IS BY THE CASCABEL" - PASAGEWAY BY THE
"AND IT TAKE'S CROSSING THE FIRST RIVER UPSTREAM, TAKING TO THE LEFT (ORIENTE SIDE)_ _ _
- CROSS THE RIO CA
TO LEFT (EAST) ON
RIO CUNGUINZA

MAP DRAWN + DOCUMENTED BY LEE ELDERS ON AN EXPEDITION IN JANUARY, 1988

RIO QUAL
BILL WRIGHT 1969
W.C. STEVENS 1972
(TIGRES)
FLYNN-ELDERS 1973
RIO VICTOR
RIO COJENZA
ELDERS 1987
RIO TUNANZA
IDE #1
MARCELO NATEMA
CASA
FLYNN ELDERS 1973
CUNGUINZA
TSR
RIO TUTANGOSA
SINGUISHA'S HOUSE
CAMINO
BURNED HOUSE
BRIDGE
WAKANI
Lee's 1st CAMP JAN 24TH & 25TH 1988
RIO MIRUMI
INZA
TERFALLS
RIO TUTANGOSA
SUCUA
30 MIN
MACAS
RIO UPANO

SHUARA WORDS & DEFINITIONS

Achuara	A Shuara tribe
Arutam	Vision
Arutam Soul	Provides immunity to murder
Arutam Wakani	Ancient soul
Canelos	Yumbo tribe
Iwanci	Demons
Jivaro	Colonial name given to the Shuara
Kakaram	Powerful ones, killers
Kakarma	Power
Musiak	Avenging soul
Namura	Sacred green stones of the Shuara
Natema	Ayahuasca, a hallucinogenic drug
Niawa	Jaguar
Pani	Boa or Anaconda
Panu	Most powerful shaman
Pasuk	Spirit helper - evil
Pener Uwisin	Curer Shaman
Shuara	Man, men, people
Specter	Spirit or soul
Tsantsa	Shrunken head
Tsentsak	Magical darts
Unta	Big man
Untsuri Shuara	A Shuara tribe
Uwisin	Shaman
Wawek Uwisin	Bewitching or evil Shaman
Wakani	Soul or spirit, spirit bird

CPSIA information can be obtained at www.ICGtesting.com
Printed in the USA
BVOW04s1514290414

352030BV00007B/214/P